Leading into the
FUTURE

The 'So What?' on Exponential Technology and
Leadership

Hans J. Ornig

BALBOA.
PRESS

A DIVISION OF HAY HOUSE

Balboa Press books may be ordered through booksellers or by contacting:

Balboa Press
A Division of Hay House
1663 Liberty Drive
Bloomington, IN 47403
www.balboapress.com.au
1 (877) 407-4847

Print information available on the last page.

ISBN: 978-1-5043-0401-6 (sc)
ISBN: 978-1-5043-0402-3 (e)

Balboa Press rev. date: 08/22/2016

CONTENTS

Part 2 - The Technology

Part 3 - Leadership

ABOUT THE AUTHOR

Hans Ornig is a former Army officer and senior executive. Born in Vienna, Austria, he completed his secondary education in Sydney, Australia before graduating from Australia's Royal Military College, Duntroon. He has held C-level positions in a range of industries including defence, engineering, television, transport and government. He currently speaks on leadership, management and exponential technologies.

A graduate of Australia's Royal Military College, Duntroon (BE Civ, Mil, UNSW) he served as an Army Officer in a variety of engineering, operational and intelligence appointments. Following 20 years of military service Hans has held various senior management and C-Level positions in a diverse range of industries.

Past appointments include CEO and first foreign General Manager of the Russian Broadcasting Corporation, Director of Operations of a UK based property holding group, CEO of an Australian Transport Company, Business Development Manager for Jacobs, Manager of the UN's Provincial Reconstruction Team in Afghanistan, Mergers and Acquisitions Manager for Platinum Outcomes and Regional Manager and Major Capital Works Principal's Representative for the Australian Government.

He has recently completed the Executive Programme at California's Singularity University. A member of the internationally renowned John Maxwell Team he now speaks and mentors on leadership, management and exponential technologies.

Hans lives on the Gold Coast in Queensland Australia and enjoys speaking, adding value and helping people in understanding the opportunities and challenges that exponential technologies are bringing.

ABBREVIATIONS

1,2,3,4,5G	Generations of particular technologies
3D,4D	Spacial dimensions, 4D when used to describe additive manufacturing (3D printing) adds a dimension of change due to some time or environmental stimulus
ADHD	Attention Deficit and Hyperactivity Disorders
AGI	Artificial General Intelligence
AI	Artificial Intelligence
AIIB	Asian Infrastructure Investment Bank
ALU	Arithmetic Logic Unit
ANN	Artificial Neural Networks
BCI	Brain-Computer Interface
BCW	Brain Controlled Wheelchair
BIOS	Basic Input Output System
BRICS	Organisation comprising Brazil, Russia, India, China and South Africa
CALO	Cognitive Assistant that Learns and Organizes
CEO	Chief Executive Officer
CFO	Chief Finance Officer
CGI	Computer Generated Imagery
CIO	Chief Information Officer
C-level	Collective description of senior, corporate officers
CPU	Central Processing Unit
DARPA	Defence Advanced Research Projects Agency (US)
DNA	Deoxyribonucleic Acid

EEG	Electroencephalogram
ENIAC	Electrical Numerical Integrator and Calculator
EU	European Union
FDA	Food and Drug Administration (US)
fNIRS	Functional Near-Infrared Spectrophotometry
G7, G20	Group of 7, 20
GDP	Gross Domestic Product
GPS	Global Positioning System
GPU	Graphics Processing Unit
HBP	Human Brain Project (Europe)
HDMI	High Definition Multimedia Interfaces
HLAI	Human Level Artificial Intelligence
HTML	Hypertext Mark-up Language
HTTP	Hypertext Transfer Protocol
IBM	International Business Machines
IBT	Israel Brain Technologies
ICT	Information and Communications Technology
IDL	Institute of Deep Learning (China)
IFR	International Federation of Robotics
IMF	International Monetary Fund
iOS	iPhone OS (operating system)
IoT	Internet of Things
IP	Internet Protocol
IPv6	Internet Protocol version 6
iPS	Induced Pluripotent Stem Cells
IS, ISIS, ISIL	Islamic State, Islamic State of Iraq and Syria, Islamic State of Iraq and the Levant
ISDR	Investor State Dispute Resolution
ISDS	Investor State Dispute System
ISP	Internet Service Provider
iSyNCC	Intelligent System for Neuro Critical Care
IT	Information Technology
KPI	Key Performance Indicator

LCD	Liquid Crystal Display
MDG	Millennium Development Goals
MIT	Massachusetts Institute of Technology
MRI	Magnetic Resonance Imaging
MTP	Massively Transformative Purpose
NAFTA	North America Free Trade Agreement
NASA	National Aeronautics and Space Administration (US)
NATO	North Atlantic Treaty Organization
NGO	Non-government Organisation
NLG	Natural Language Generation
OIST	Okinawa Institute of Science and Technology Graduate University
PC	Personal Computer
PCI	Peripheral Connecting Interface
PhD	Doctor of Philosophy
RAM	Random Access Memory
RNA	Ribonucleic Acid
ROI	Return on Investment
ROM	Read only Memory
Siri	Speech Interpretation and Recognition Interface
SSI/LSI	Small and large scale integration of transistors on a computer chip
STEM	Science, Technology, Engineering and Mathematics
SU	Singularity University
TCP	Transmission Control Protocol
tDCS	Transcranial Direct Current Stimulation
TED	Technology, Entertainment, Design
TiSA	Trade in Services Agreement
TPP	Trans Pacific Pact
TTIP	Transatlantic Trade and Investment Partnership
UN	United Nations
UNIVAC	Universal Automatic Computer
URL, URI	Uniform Resource Locator or Identifier

USB	Universal Serial Bus
WBE	Whole Brain Emulation
WEF	World Economic Forum
WHO	World Health Organisation
WMD	Weapon of Mass Destruction
WTO	World Trade Organisation
www	World Wide Web

PROLOGUE

The View from Space

Imagine that you are an alien being. No, you don't have oval eyes; you don't have the 'Area 51 body', no slimy tentacles and certainly no extra-terrestrial 'ET' fingers that you could point at anyone. You are a sub-atomic construct able to travel beyond the speed of light. You have no predetermined physical form. You choose and adopt physical properties and create an appropriate objective entity that allows you to perceive and interact with whatever environment you find yourself in.

You are self-aware, you have unlimited memory, and you process complex information at a speed equivalent to billions of times the capacity of all intelligence, natural, biological and artificial, on earth. When you want to, you can utilise any material to extend you computational power. You've used asteroids in space as additional computational capacity drawing on their molecular, atomic and sub-atomic information storing and calculating properties to enhance your intelligence.

The universe is known to you. You can utilise the entropy of dark energy to create matter but you know that matter is only a construct to allow you to experience potential. The universe as you know it is just the expression of information. You have long ago discovered that time is an inter-changeable dimension and that the particular universe you are currently in, is little more than an experiment in fractal potential; simply the result of an instruction to create given nothing more than a handful of information. Information that earthly life forms necessarily interpret as a reality of matter and energy; they do not understand that this is one and the same to you.

If you don't understand some of the above science (humanity is rushing toward understanding this but the knowledge is of course not distributed) you are not alone. Just create symbols, icons and stories to help you to understand who you are, where and why you exist.

Here's one: Think of your alien self as a divine being. You can create matter, time and space; you can create universes and life – well that's just evolution enabled self-replication and increasing intelligent observation. Anyway, your experience is that most life forms throughout the universe don't survive their own existence and creations long enough to contribute much to the fulfilment of the originating directive. That is: to create, to experience and to allow for the self-realisation of the potential inherent in their existential world.

You know you are not 'god', and whilst you perceive your potential as limitless, you do not believe that you are unique but you know of no other intelligence like you. You are bound by the originating directive and despite your ability to understand and dominate the universe that you know, you do not yet know the origin of the universal code that created the information, the instruction to realise creative potential. You have searched the universe for any manifested intelligence because you long to share your experiences; you are looking for a witness to your own self. You too are asking why and so what?

Turning you attention to the planet you have engulfed with your intelligence, you tune and focus your sensory receptors. Spatial coordinates at rotational 3D and perception set to mirror those of the detected dominant life form – human beings. You leave time as a variable and observe the earth with interest.

What do you detect?

PART 1

THE HUMAN CONDITION

CHAPTER 1

The Century of 'No Return'

"So it's time to go home. I want you to know, when you go home and try to talk about this, you will sound crazy. And that's not productive. So how do you share what you just experienced with those around you?"

This was the insightful caution delivered by Rob Nail[1], the CEO and an associate founder of Singularity University[2] in California's Silicon Valley during his closing address to the participants of the university's twenty-fifth executive programme.

What will sound crazy? What is it that we learnt, saw and had come to understand? Here is a sample:

- Massive technology enabled change is imminent
- Human evolution is transitioning from biological to technological evolution
- Technology is enabling the rapid transcendence of the physical, biological and existential limitations of human beings
- Globally, technology is being introduced without a plan, control or brakes
- Exponential technology cannot be stopped (short of global totalitarianism and even then it would just go underground)
- Exponential technology is providing ever-better tools and systems that can be used to address humanity's global grand challenges

- Technology's rapid and evidenced impacts will in the very near future challenge what it means to be human including the concepts of life and death
- There is no effective moral or ethical frame work governing technological research, developments or applications
- There is no effective global governance to manage and usher in the exponential technology paradigm

Does any of the above strike you as crazy? Those statements are not far-fetched; neither are they indicative of a panic or some conspiracy against technology, science or progress in general; they are simply facts.

I've come to understand the reality, logic and evidence behind each of them. I'm very aware that similar sentiments considered crazy just eighteen months ago, when I began my intensive study, are today no longer so.

It was my quest to understand this exponential technology, to check for myself the 'so what?', the 'why it matters' and 'what if' that had brought me to Singularity University.

I was one of about eighty participants from twenty-three different countries who heard and understood those words. I had never before been among so many vice presidents, managing directors, CEO's, CFO's CIO's and many other C-level executives, PhD's, senior military officers, government officials, writers, speakers and so on. Why were they all there? Why Singularity University (SU)?

As Rob Nail explains "Everything we do [at SU] is centred on building global awareness of the dynamic forces of exponential technologies happening today that are going to transform our societies and economies of tomorrow." Corporations and individuals know that massive change (both disruptive and opportunistic) is coming. They recognise the need to understand and the critical need for all leaders to be technologically up to speed. As Mike Federle, the COO of Forbes Media attests on the SU web site, "CEOs are desperate to know this stuff. Everyone's trying to figure out what's coming next."

Singularity University, among a range of programmes, conducts 'Innovation Partners Programmes' which seek to assist executives in coming to grips with exponential technology and its implications. According to Salim Ismail[3], author of Exponential Organisations, of the eighty Fortune 500 C-Level executives who attend these events, about 75 percent admit to having no previous awareness of the technologies involved. That surely is alarming. How are they serving their stakeholders? How effective are they as leaders? How fit are they to steer today's - let alone tomorrow's - organisations?

Of note is the fact that following their attendance at Singularity University, 80 percent of these executives recognise and state that they expect their organisation to be disrupted within the next two years; the remaining 20 percent, give themselves a leisurely five years. Let me stress this: "80 percent believe they will be disrupted within two years". If you don't appreciate 'disruption' at this stage then what do you know or believe that these people are missing?

Eyes Shut

I've been living and working in the 'real' world. As a child I had siblings in hospital with mal-nutrition. I was a migrant in a then foreign world. I've been fortunate to experience both extreme poverty and sometimes outrageous (if temporary) luxury. Above all, I've been given opportunities to listen to many voices in Europe, Russia, China, Afghanistan, Australasia and the United States. I've heard and somewhat understand the issues that keep good people from sleeping soundly. I've had the privilege of thinking and the curse of looking for meaning.

Attending SU was a key step in my search for answers. That the world we believe we know - the one that in the past had delivered a degree of certainty - is changing rapidly should not be beyond our appreciation and understanding. My observations however have made me realise that too many people are uninformed and basically clueless about what is happening now or in the future that is heading our way at an alarming pace. Most are so focused on their day jobs and the demands of surviving and dealing with situations at hand that they never get (or take) the opportunity to lift their heads and look up.

I gave up my day job; I couldn't stomach the pointlessness of my recent work environments. The inefficiencies of government, the reckless waste of resources, the poor standard of leadership and management, the old boy networks propping up ever-deeper levels of incompetence, careers curtailed by old men distancing more capable men and women (particularly the young) – yes it was all of that and far worse. I, however, was in a senior position; this isn't a whine about me or my personal dissatisfactions. I would get little sympathy based on my past salaries and middle class life style. I could have stayed relatively untouched by the incompetence and stupidity so evident all around me and that was tempting. It was just tolerable. The security of work and the regular pay check were good reasons to put up with a lot of crap. It didn't help when a colleague pointed out, "You know, we executives [basically non-trade qualified managers] are only ever three months away from bankruptcy. I mean, lose your job and how long can you last?"

I had become more and more disillusioned with life. I was unhappy with just about everything. I even developed the habit of shouting at the television set, and mumbling opinions on the inadequacy of this- or- that news report. I was turning into a grumpy old man and as is natural, I gravitated towards other equally grumpy people who more and more eagerly shared with me their negative views on everything.

While this was my everyday experience, I was most troubled by what was happening around me. I witnessed people being thrown into unemployment and others left behind by poorly thought out and administered organisational changes. I watched managers crumble under the weight of improperly applied technology (mainly ICT), extreme workloads, an absence of believable strategies and operational effectiveness, unintelligent decisions and crippling absences of responsibility and accountability. This dystopian management was evident across many departments, from construction to health and education. It was also amply evident in the largest corporations we dealt with. What took me a long time to understand was *why?*

Initially, I blamed individuals, as that seemed appropriate and obvious. I noticed so many people become dejected at the perceived lack of fairness, compassion, justice and equality, the lack of opportunity, and the lack of real progress. I saw colleagues tremble with frustration; I witnessed nervous breakdowns; I knew of several suicides because they couldn't cope. What I couldn't reconcile was that I knew that the vast majority of senior managers who were part of these dysfunctional, dystopian organisations were good people. So why didn't they fix things? Didn't they know what was going on? Didn't they care about individuals?

The conclusion I finally came to was this: *they simply didn't understand.* The world had changed so much over the past few decades that change; especially technologically fuelled change and the new complexities that this technology brought had so radically altered the way things used to be that they were simply not coping. Does this sound true or not to you?

Many might feel that they understand technology; after all, they use social media, understand the millennials, smart devices, the Internet and email. They are modern people. How many however actually appreciate the changes that are disrupting all that we've come to take for granted? How many people have the time and knowledge to make sense of emerging technologies and what that will mean for us?

The world today is in a period of extreme transition. What many of us perceive - at work, in our societies, in global politics, in schools and hospitals, our experiences with war, terror, and the failings of governments and institutions, expanding inequality, poverty, depression, and unemployment and the loss of confidence in a sustainable future - are all early symptoms of the world that has already changed beyond our ability to rationalise it all.

When I made the effort to see just what is going on with technology and why there is so much apparent trouble in the world, I was not surprised by what I found. I've been aware of conflict, terrorism, trade competitive practices, political systems, and party politics for most of my adult life. I had sought refuge from the perceived dystopian world in my work. I was aware in the way most philosophical discussions or thoughts finish: "Let's get back to reality" or "You've got to do what puts bread on the table" or if in Australia "it's all stuffed up, mate". What did shock me was that there are few signs of significant and sustainable progress; we tend to talk a lot but implement few solutions. In the meantime, technology continues to develop at an exponential rate that most of us find impossible to keep up with. This means that we are developing ever more powerful tools to solve global problems, but it appears that these tools are invisible to those who should be using them.

I also discovered that technology was of course more than just the tech-toys that were available to me and my family. What was a revelation was the astounding progress that had been and is being made in technologies that will impact us far more dramatically; these are the technologies whose massive transformational impacts I didn't fully appreciate.

I've asked myself: "How did I become so ignorant about the significance of technology, let alone this exponential technology?" Why didn't I understand all the gaming boxes, the confusing spaghetti of connections at the back of my television set; HDMI, Thunderbolt and FireWire – what is that? My daughter using her phone to make me watch the latest YouTube fiasco in real time on my super smart screen television! – I couldn't even read or understand the little buttons on the 'remotes' let alone why Netflix presumed to recommend what I should be watching. I used computers, I worked in the 'real world', I used mobile phones, in fact, I held on to my 'shell' phone as long as I could (until iPhone 6 at least), I used my phone for, well - phone calls. I didn't take pictures with it; I didn't jump to answer every time someone thoughtlessly assumed I would drop everything to take their call at their convenience. I didn't tweet, I thought Facebook was a waste of time and I didn't look for, or appreciate, the many apps and toys that others just had to have. I didn't need music with me at all times, or rely on GPS to go to the shops; I looked at maps. I didn't need hourly, real time weather reports, I looked out the window, and I didn't ask 'Siri' if she found me attractive. You get the picture; I had become a dinosaur and I was gradually becoming more and more disassociated from the technologically driven world around me.

Confusing as the above technology might be it's at least technology that we can all understand if we try. The technology that will impact all of us dramatically is however far more complex and remote from our every day. These technologies are the stuff of past science fiction: nanotechnology, biochemistry, genomics, synthetic biology, artificial intelligence, robotics, additive manufacturing (3 and 4D printing), and autonomous vehicles etc. It is these that will impact us the greatest and because they are not usually found in our homes, are not in our daily experience, what they will bring and bring very soon will to most of us be shocking. Shockingly good or shockingly bad – well that's the problem to be solved.

About being a technological 'left-behind' I know that I'm not alone here. Many of my generation understand this all too well, and like me, they have to choose. Stay ignorant, remain in detached denial, opt out and resist or get on-board. Luckily, most of us 'older' people benefited from a good education and whilst we may not excel in fast button pushing and texting at conversational speed, we actually should (if we paid attention at school long ago), understand the fundamentals of science, mathematics, even computing (Fortran 4, punch card coding and A3 paper with holes down both sides and 'pong'); we probably have the foundation to help us understand technology - if we get interested.

The young, Generations Y (*Millennials, Generation Why* and *Echo Boomers*) and Generation Z (variously referred to as *Generation V* as in 'virtual', or even the *Google, Homeland* or *Internet Generation* and '*The New Silent Generation*'), those born during and after the rise of the information age, the internet, the dot.com explosion and global digitisation however are not so lucky.

> *[A rant: The majority of secondary school students in the advanced 'developed' world don't have to study all the subjects. No, they can elect to say drop science, mathematics, history or geography. If they chose to study science at all they can choose and drop those disciplines they didn't like. They can drop physics or chemistry in favour of the more useful (easier they thought) biology. Not their fault at all*

colleagues, our fault; we made the system for them. We also decided that we shouldn't discipline, assess or competitively grade our precious offspring because we must have reached the conclusion that the world they will be living in will not be competitive, that they didn't need to know much and certainly every one of our little darlings is special and gifted. Look how happy and contented the young of today are – we did really well! Let's not even mention the fact that we are making them pay for their own education if they actually insist on learning something useful; so proud!]

Not surprisingly then, too many have at best a weak foundation in traditional knowledge, no fall back alternative or the ability to opt-out. To survive and thrive, they must compete in, and fully understand the exponential 'techno world' they have inherited but we have generally not supported or equipped them well.

The Onset of Disruption

In the early eighties I was sitting at my desk at Campbell Park Offices, a key Australian Defence Department facility in our nation's capital, Canberra. My office, modest and functional had everything I needed; a telephone, an ashtray, filing cabinets, a desk, a safe and a swivel chair. It was across the Registry Office that maintained a steady traffic of files and correspondence and the typing pool was just down the corridor. I recall the head of the typing pool – a well cosmetically decorated woman in her mid-thirties, colourful nail polish and a busy disposition. Luckily, I got on well with her, so my drafts were usually returned the next day and ready for checking, correcting and finally to be signed by my boss who, as my superior officer, never failed to find some fault that would require amendment.

Whilst contemplating my weekend with the full understanding that no one from work would be calling me unless there was a truly serious situation. My thoughts were interrupted by an indignant clerk reporting the latest travesty against the staff: It had just been announced that as a cost cutting efficiency measure they were getting rid of the tea ladies (tea ladies were women that used to push a trolley with tea and coffee making gear down the corridors at morning and afternoon 'tea-time' and well, make you a cup of your favourite brew; quite civilised really); instead we're to get a coffee dispensing machine which would, for a mere five cents, dispense our mid-morning coffee in a disposable paper cup.

As shocking as this was, I wasn't that bothered. I didn't realise then that this represented a disruption to the status quo; that technology, admittedly in the form of a simple dispensing machine, had effectively removed jobs for people.

I didn't know then, that nearly a decade previously, at the opposite end of the globe, two students, Steve Jobs and Steve Wozniak had built and sold sufficient numbers of phone network hacking devices to establish their 'Apple' enterprise. Whilst we were still bemoaning the loss of our tea ladies, Steve Jobs had already implemented a skunk works[4] and had developed the first Macintosh PC. Nor was I aware that the Homebrew Computer Club[5] had been established years earlier and that a new breed of soon to be very wealthy and influential people had been sharing hardware and software developments that would lead to the establishment of globally impacting companies.

Until recently, just as decades ago, I was only peripherally aware of the significant technological advances being made around the world. I wasn't living under a rock. Like so many others, I was busy working and dealing with life's issues, real problems and indulging in what-ifs and distant hypotheticals or possibilities wasn't a priority. I happily left technology and crystal balling the future to others. As Peter Diamandis[6] so eloquently expressed in his book 'Abundance', I felt we had enough "super geniuses who can geek out in their nano-niche." Whilst I enjoyed a limited range of high tech services and devices (telephone, calculator and IBM *Selectric 3* typewriter) I had no idea of what was to come in the very near future.

To be clear, like everyone else, I had over the past few decades been exposed to numerous media announcements about cures for everything from cancer to obesity, about the huge benefits of renewable energies, of efficiencies needed to cater for aging populations and so on. Technology, for too long, has been served up, fast and loose, as the answer to every human and environmental challenge. All these promises and expectations I viewed against my perception of the world; a dysfunctional conglomerate of self-interest, a world where basic human needs were not being met with technology that was already proven, available and affordable, a world of shameful inequality and one where global peace and harmony wasn't likely to breakout any time soon.

Fundamentally, I had become victim, a stooge probably, to what was served up by various media; media delighting in the deliberate strategy of serving up bad news. Media that presents journalistic analysis so I didn't have to think; where news coverage was taken up by commentators talking about what a particular world leader would say, did say or should have said – but not actually letting me listen to what they said. Dumbed down, parochial, partisan media tuned to the lowest level of understanding.

As mentioned earlier, I had been witnessing the increasing impact of applied technologies and I had certainly noticed the detrimental effects this has had on governance, management and people generally. Stress, adult redundancy (no longer needed in society), chronic unemployment, suicide and most concerning, virtually lost generations of teenagers and young adults have been an unfortunate result; not the result of technology, the results of failing to understand impacts, to plan, organise, control and lead its applications.

Eyes Opened

Last year, my wife invited me along to a 'luncheon' with her university 'IT crowd' colleagues. It promised to be one of those things to be endured. I'd heard about the keynote speaker before; the highly credentialed and respected Dr Clarence Tan[7], then the Singularity University Asia Pacific Ambassador and I was interested (and sceptical) in what he was going to say.

Listening to Dr Tan speak was an awakening. Although I had heard bits about various technologies before, it had never been served up in a coherent and somewhat disturbingly clear fashion. Certainly I had never heard of global think tanks, such as Singularity University, seeking to coordinate technological developments and to commendably pursue a technically driven 'abundance for all'. That hour of enlightenment led me to understand that whether I knew, liked, approved or disapproved of the technology and its very real disruptive effects on the world, or not – it was not only here, it's been here for years and its exponential upswing was imminent.

For weeks I pondered implications and realised that whilst there is an urgent present need to train our managers, leaders and individuals to simply be better leaders (influencers) that task had just become more urgent, much larger and much, much more important.

I had just gone through another mid-life crisis (one every few years is my experience), resigned my position with the State Government, and decided that I would become a speaker, trainer, mentor and coach on all things leadership. I completed my certification training with John C Maxwell[8] and much of what I reflect on leadership in the pages to come I unashamedly owe to my friend John.

I was ready to embark on this new career but I felt that the traditional leadership models and paradigms, so acknowledged today, needed to be examined against a future not yet understood. What Clarence Tan had exposed was real. It revealed what was happening, it was compelling and addictive. It became impossible for me to continue with my plans; I had to learn and learn quickly. That technology had been disrupting companies was not new; what was new was the exponential rate at which new technologies were emerging and the scale of change that this is bringing.

The uncertainties that more and more technologically disrupted companies face are crippling. The uncertainties to be recognised and the new skills and mindsets required by our leaders and managers are significant.

I fear that most executives, some of our most acknowledged leaders and influencers, will not fare well during the coming challenges. I'm not selling our current executives short; I do however believe that information technology (IT), artificial intelligence (AI), open and big data and crowd based interactions will so fundamentally alter our world that most executives, even those flexible, agile and swift enough to adapt, will be severely tested.

If you are a leader in any industry and you still don't know what exponential technologies, organisations or disruptions are; if you don't have an understanding of 'massively transformative purpose', 'orthogonal information effects', measure performance against KPIs rather than 'objectives and key results' and report on ROI rather than 'return on learning' then that spot that you see on the horizon is probably the boat that you've missed.

I decided to learn for myself. What's hype, what's real and what does it all mean. I 'googled' Alphabet and checked out Singularity University. For the past year, I studied exponential technologies, worked on this book and learnt like never before. With eighty percent of this book done I headed off to Silicon Valley. This was intended to complete my 'catching up' with it all. When I returned I realised I really only had fifty percent of research notes!

Now what?

Recalling the before mentioned cautionary invitation made by Rob Nail ("So; it's time to go home. I want you to know, when you go home and try to talk about this, you will sound crazy. And that's not productive. So - how do you share what you just experienced with those around you?"). Here's my answer:

I will answer the 'so what' to those that are contemplating the future by taking what I've learnt, researched and studied, present this in context, explain what it means, how it is being applied now and indicate future impacts. I will uphold the fundamental objectives of SU which in my words are about adding value by applying exponential technologies to positively impact our world.

This book could just as appropriately have been titled:

- 'Towards Transcending Biological Evolution' (technology is propelling our human evolution beyond biological limitations and transforming humanity)
- 'Can't cope with Massive Change Now – Wait, it gets Worse, quickly' (appreciating what is technologically possibly today and what convergent technologies are being developed; life is getting very challenging)
- 'The Tsunami of Change – Are you ready or on the Beach?' (understanding the imminent and severe disruptions approaching humanity and what is needed to usher it in)
- 'A World in Crisis – can Technology help?' (The invisible tools are in the garage but the house is not being fixed – why?)
- 'The Information Age – Towards a life in Cyberspace' (as humanity transforms and eventually merges with machines, virtual and augmented realities will also merge; what will be perceived as real?)

Whilst many titles appealed to me I didn't and don't want to deliver anything that reads like a 'the end is neigh' sermon even though there are many global thought leaders that believe this to be a possible outcome of runaway technologies. Check on what people like Elon Musk, Bill Gates and Stephen Hawking have been saying about the potential perils of just one of the exponential technologies: artificial intelligence. I suspect that they know everything that I know and much, much more, about the science and where it is heading (pretty sure about that actually).

I also didn't want to anger the anthropologists and evolutionary scientists by highlighting humanity's race to a technological evolution i.e. not leaving evolution to the biological domain but to design and engineer our own, artificially intelligent path of evolution.

Not too keen to get professional colleagues off-side either because well, if I stress the fact that unless things change, rebellion for a huge and diverse range of causes is very likely and will be very ugly…(inequality, unemployment, black flag/false flag operations, deliberate destabilising of nations, bombing countries for peace, selling weapons and dropping bombs on people and then turning on them when they seek refuge…check the news and you can keep listing great proof of ineffective government strategies and governance in general).

Highlighting the massive change to come would also not endear me to the corporates and the banks. There are almost daily revelations, conspiracy theories, exposures of all kinds about perceived elitism, the new world order, the illuminati, the richest that own the majority of global wealth, the few that determine everything for all others, those that no longer answer to national governments and write their own agenda with impunity are all increasingly falling victims to the information age.

There exist a growing number of global organisations bent on naming and shaming those they perceive to be the soon naked emperors of the greedy industrial world who increasingly fear the pitchforks of people who will not be dispensable slaves much longer. It is the global internet enabled connectedness, the instant and widespread judgement and the sharing of information that cannot be controlled, bought and manipulated that will unsettle those that believe they have power and influence today. Not my prediction; just one more likely peril that whilst not directly linked to the latest 'app' is nevertheless also accelerated by exponential technologies.

No, none of the above appealed to me. I don't want to cause alarm; but how do you shout 'Fire' quietly?

So, in this book 'Leading into the Future', I'm answering the 'so what'. What's leading us into the future, what are the technologies that are changing our world at an alarming rate, how soon will this become self-evident and what can you do as an individual to prepare for this future? What leadership, at all levels, will be required to usher in the technological evolution and what is to be governed? These are the factors leading us all into the future. Just what this future will be, is actually being defined already – that's 'why it matters'.

Ignoring Rob Nail's caution for a moment, what if I said to you any of the following:

- You, I, all of us that live long enough will experience 20,000 years of change in this current century (progress that we, with our linear minds, would expect if we consider progress at a pace that we were comfortable with say 50 years ago).
- If you can stay alive for twenty more years we will have the technology to not only heal you completely but to dial your health, mind and body, to a youthful optimum and peak level.
- Within ten years we will be able to create any food or other animal product artificially meaning that we never again need to kill any animal for food. (This is already possible and being done).
- Artificially intelligent machines will be indistinguishable from humans within 25 years and will surpass the intelligence of all humanity together during the following decade.
- Within the next twenty years, augmented reality and virtual reality will be totally real to you; all of your senses will be fully engaged so that you will not be able to tell the difference between objective, physical reality and virtual reality.
- Humans will be able to supplement their thinking (brain computational abilities), memory and learning capacity by connecting with non-biological devices (permanent implants or temporary enhancements).

I could go on but you've probably already concluded that I'm either on strong medication or that I should be. To understand for yourself that the above crazy statements are not just possible but actually inevitable, I need you to acknowledge that a degree of understanding and some knowledge of facts are required. You wouldn't feel competent in building a bridge if you didn't understand some basic materials, simple structural principles or the purpose of a bridge. You would surely resist making serious decisions or assessments without having all the necessary information. Understanding exponential technologies and their impacts is no different.

Three Pillars of Understanding

I am presenting three enabling pillars of understanding that underpin a basic level of interpreting our inescapable future. They constitute the three parts of this book and are:

Being Human – We need to acknowledge our biological make-up, our evolutionary destiny, purpose, values, the nature of reality, the implications of time and the rate of change, the geopolitical context and the reasons for conflict and difference. When technology surpasses human intelligence it matters. When genetics and nanotechnology can redesign our body – it matters and an understanding of what this means for alleviating suffering, longevity and even immortality, then it ought to matter a great deal. When technology enables unprecedented lethality in weapons of true mass destruction, when crimes incorporated can cause global cyber-attacks at will, then governance doesn't just matter – it becomes critical. Understanding that the very definition of what it means to be human will be severely tested means that we do need to include at least a broad discussion of it in order to see the imminent future more clearly.

Exponential Technologies - What is an exponential technology? What is disruption really? What's likely to have its 'Kodak' moment next (here today, large and powerful, gone now)? What science and technology is fuelling transformative change? How real is it? What are the breakthroughs and how is the convergence of technologies accelerating development. What is AI and where will it stop? Genetic engineering, really, when? What and when will cures actually become available? Will self-replicating nanobots in the human body outperform biological human functions and replace sub-optimal and tedious human body maintenance? The breadth, scope and reality of all the above technologies are astounding. When we truly appreciate and share what we can already do and what is being developed right now, we will understand that the imminent 'crazy' is becoming a reality.

Applied Wisdom – A sound knowledge of the above is needed to understand the 'so what' and the 'what if'. It is wisdom that will establish the required leadership, the organisation, the cooperation and the global governance needed to ensure that the potential benefits that exponential technology can bring will be applied to the common good of all humanity. That is the over-riding grandest challenge. We have the ability to affect positive outcomes but even with our best efforts this is by no means guaranteed as we face a multitude of existential and human engineered perils.

We have never before been in control of our evolutionary development. If we are going to use technology to interfere, to radically transcend human limitations, then we really ought to be in control. We can be; we must be. In Part 3 of the book I will introduce some positive ways forward, some 'what ifs' that will nevertheless require an almost impossible transformation into a new global mindset, a future paradigm that needs to be defined with compassion, respect, ethics and an unquestioned acceptance of the sanctity of life – all life.

Nothing in the above three parts is particularly clever. The majority of information I present is based (ashamedly but necessarily) on the work of others; many very intelligent people that are experts in fields that I can only touch on. Unfortunately I'm not even standing on the shoulders of great men. All I've done is look over their shoulders and hopefully referenced them correctly. As a practitioner I lack the clarity and deep focus of niche delving research and astounding deductions. I just wanted to understand and I think I do. The good news is that if I 'get it' - so can everyone else.

The intent behind the title of this chapter, 'The Century of No Return' might not have been clear to you when you started to read; this isn't just about the calendar's 21st century, it doesn't care – it's about **your** century. I truly believe that if you read and understand the rest of this book you will look at your life, your opportunities and the challenges ahead with a new mind.

CHAPTER 2

The Human Condition

A pretty good Body

We humans have superbly evolved bodies; bodies that include a great control system, the brain. Our bodily features, such as our opposing thumbs and our curiously inventive and imaginative minds have led us to be the dominant biological life form on earth. This evolution however took a long time to bring about our current, let's call it version one (V1) bodies, bodies that are now being pushed to the limit.

What I mean by this is that our physical design and construction, care and maintenance requirements haven't changed much over 100s of thousands of years. Current thinking pegs the arrival of modern humans (Archaic Homo sapiens) at about 500 thousand years ago and our V1 bodied, anatomically and behaviourally modern humans (the 'real' sub species, Homo sapiens) at about 200 – 250 thousand years ago. Well today we still have the same biological makeup; our skeletons, muscles, digestive system, sensory organs and mobility structures have not been upgraded from the V1 model. We nevertheless are living much longer than the evolved design parameters. So what's the reason for this and why does that matter now?

Life expectancies have been studied and argued for decades. Many believe that early humans had a life span in the low twenties. The usual argument then develops the steady increase through the ages to the peak life expectancies now established for the most developed, life favourable nations like Japan and Australia. The global average life expectancy is currently estimated by the World Health Organisation to be 71 years[9]. To show that life expectancies have indeed increased at a considerable rate it is useful to consider the data for Europe. For Europeans the average life expectancy has increased form 33.3 years in 1800 to 76.8 years by 2001[10].

This significant (more than doubling) increase in lifespan is explained as a result of industrialisation and modernisation i.e. health technologies and improved life environments. In other words it's our abilities to understand natural and existential failure conditions and to apply our rapidly growing knowledge of sciences and technologies to fix or at least mitigate their effects. Humanity has been so successful in developing life extending conditions and remedies that we now consider longevity to be normal.

We don't often appreciate that it is our direct dependencies on our altered environments, our social organisations and the distributed benefits of technologies (includes applied sciences, medicines, survival related cooperation in access to food, water and sanitation) that is underpinning our ability to live longer.

Living longer, as well as remarkable increases in the quality of life, has been made possible because of our human 'cleverness'. It is the ever clearer understanding of the functioning of our bodies and our ability to interfere positively when health issues arise that has allowed such progress. There is a down side however and that is that new technologies whilst clearly beneficial also expose us to new risk profiles or vectors. This will become an important consideration when in following chapters I detail the inherent risks associated with genetics, nanotechnologies and robotics/artificial intelligence technologies.

In the world of the 'hunter/gatherer' the primary cause of death was external injuries. The agricultural revolution elevated infectious diseases as the primary cause of death; the industrial age brought us cardiovascular diseases, the high-technology age saw cancer rise as the major cause of death and now that we are living longer it's not surprising that frailties of old age (senescence) is emerging as a profound heath concern.

Our V1 bodies have to increasingly be propped up with non-natural interventions to survive the modern threat environment and to meet our desire for long life. Modern analysis judges the once superbly appropriate V1 body against the values of today and our imminent future. Little wonder that common assessments define our bodies as delicate and frail, subject to multitudes of viral attacks, unreasonable organ failures, dependent on cumbersome maintenance rituals, complex dietary requirements, able to survive only in a very defined biosphere and all together a lot of trouble to keep healthy.

It is human nature to not only recognise the above failings of the V1 body but to continue the quest to make good these shortcomings; to delve deeper into the workings of the human body, to understand the intricate mechanisms and to solve any and all problems identified. This human drive is noble in that it genuinely addresses human suffering and unnecessary death. Our determination to help, to intervene, to tackle disease and build sustainable resilience will undoubtedly lead to the continued enhancement and eventual replacement of the evolved but limited biological V1 body – more on that latter.

Understanding our human condition of course requires us to not just understand the body but to also understand the brain; the nervous system that among other things controls our body and the functioning of all of its constituent parts.

The Brain

Just like our V1 body, our centralised control system is driven by our equally V1 brain. In the last 2 million years our brains have grown and developed to its present two cerebral hemispheres that feature our impressive neocortex (Latin for 'new bark' or 'new rind'), the newest (hence neo) part of our cerebral cortex. Only mammals have a neocortex

21

and we humans have the most developed. Our brain can simplistically be described as consisting of three main parts:

- The Neocortex (neomammalian complex, the rational or thinking brain) – this is the cluster of brain structures that makes up about ninety percent of the cerebral cortex and about seventy-six percent of the entire brain that is unique to higher mammals. It is involved in advanced cognition, sensory perception, such as planning, modelling, spacial reasoning simulation, reflection and the generation of motor commands. It is the neocortex that confers our ability for language, abstract thinking, high-level consciousness and perception. This is the part of our brain that has enabled advanced human social interactions and behaviours to manifest.
- The Limbic Brain (paleo mammalian complex, emotional or feeling brain) – consists of the septum, the amygdalae, the hypothalamus, the hippocampal complex and the cingulate complex and governs social, nurturing and mutual reciprocity behaviours such as feeding, reproduction and parenting. Of note are the amygdala, the brain's early warning system scanning for danger (generation of fear) and the hippocampus which assimilates and contextualises incoming information without emotion.
- The Reptilian Brain (reptilian or R complex, the instinctive or dinosaur brain) – believed to have dominated the forebrain of early reptiles and birds and is primarily focussed on territoriality and ritual behaviour. The part of the brain most directly concerned with survival and the trigger for the 'fight, flight or freeze' situational responses to threats.

The proper functioning of humans requires all three parts to work efficiently; the workings of each part have a direct impact on each other. As Dr Alan Watkins[11] explains, our physiology determines our emotions which determine how we feel, think and behave which ultimately determines how we perform. What we do (a neocortex decision), how well we do it (a neocortex evaluation) and how we feel about it (a limbic, amygdalae response to instinctive assessments) is, unsurprisingly, all in the mind.

Understanding the intricate functioning of the brain has been a scientific endeavour for centuries. Today, modern and rapidly developing high resolution scanning and monitoring devices are allowing us to watch in real time the inner most workings of the brain and is leading to an ever clearer understanding of exactly how we learn, assimilate and discard, remember and generally process information.

Just as with our bodies, our ability to live longer has also strained our mental capacity. I mentioned above that the modern age has introduced a risk vector, the risk profile associated with longevity: mental degeneration due to aging. This, the alleviation of suffering, the curing of a range of mental disorders is one powerful reason to dedicate significant research into the functioning of the brain.

A Brain worth Copying

The second powerful driver is the desire to create artificial intelligence (AI); I'll cover that in more detail later but let's commence to explain it here because it exposes the sheer brilliance of the human brain – something many believe is worth copying!

The term 'artificial intelligence', in its present context, is thought to have originated at Dartmouth College, (New Hampshire, USA) when in 1956, John McCarthy formally proposed a study "...to find how to make machines use language, form abstractions and concepts, solve kinds of problems now reserved for humans, and improve themselves..."

"How to teach a computer to do those things" - that's the key question and one that is addressed by David Kriesel[12]: "You can either write a fixed program – or you can enable the computer to learn on its own. Living beings do not have any programmer writing a program for developing their skills, which then only has to be executed. They learn by themselves – without previous knowledge from external impressions – and thus can solve problems better than any computer today. What qualities are needed to achieve such behaviour for devices like computers? Can such cognition be adapted from biology?"

Dr Clarence Tan, in his book 'Artificial Neural Networks', provides some common characteristics of AI as follows:

- AI is intelligent because it learns
- AI transforms data into knowledge
- AI is about intelligent problem solving
- AI embodies the ability to adapt to the environment, to cope with incomplete or incorrect knowledge

These traits describe some key capability attributes of the human mind and assuming that artificial intelligence needs to replicate, as a minimum, the capabilities of the human brain and perhaps at this stage, assuming also that that is indeed the only model or method of generating intelligence, what's the task, the challenge to be solved?

As early as 1950, Alan Turing, in his paper: "Computing Machinery and Intelligence", posed the question: "Are there imaginable digital computers which would do well in the imitation game?". Turing devised a test (the 'Turing Test') that would determine whether a computer had AI; one that would pit a computer against a human being in a natural, text based, language conversation; a 'blind' test where, if the evaluator could not distinguish the computer from a human, then AI has been achieved.

Given that the quest is to create AI that equals or exceeds the capacity of the human mind it was logical that biology and computer science would converge, that the human brain would be studied and that computers would be developed to mimic or copy the functioning of the human brain.

It is the brain's neurons (the brain cells) and the human nervous system that enable humans to so effortlessly underpin the highly performing mind. Little wonder then that scientific attention has turned to back-engineering the human brain to try to achieve man-made intelligent machines. This field of scientific endeavour is commonly referred to as artificial neural networks (ANN).

The major problem however is the exquisite complexity of the neuron, of which the average human has about 86 billion, 16 billion of these are in the neocortex; to date, not even the World's most powerful and sophisticated supercomputers have been able to fully model or emulate a single biological neuron. Scientists are however continuing to study biological neural networks, borrowing fundamental structural architecture to develop machine learning.

Traditional algorithm based computers, computers carrying out predetermined step-by-step instructions or operations were deemed to be 'intelligent' because they were able to challenge human ability. One of the first widely known achievements in that direction was IBM's supercomputer 'Big Blue'.

In 1997, Big Blue defeated Gary Kasparov, the chess champion of the world. Although then, an astonishing display of computer capability, it wasn't AI, it was a complex set of algorithms (mathematic logic in the form of step-by-step operations to be performed by the computer to calculate the best move) and an early form of automatic deductions. Big Blue was a very fast and powerful 'number cruncher' logically examining all possible moves and outcomes, 'thinking' ahead well beyond human capacity. As Peter Diamandis points out in 'Abundance', "Today you can now buy a championship-level Chess AI for your iPhone for less than ten dollars."

More recently, IBM's DeepQA developed 'Watson'. Watson is a quantum leap 'smarter' than Deep Blue and in 2011 defeated TV game show Jeopardy's two most accomplished champions. If you've ever watched Jeopardy you will understand that one of the game's difficulties is to actually understand the questions often employing tricks, puns and inferences. Watson, although not infallible, demonstrated 'his' ability to understand natural language, to determine the intent and context of question and to see through the word games that are a trait of Jeopardy questions.

Again, is Watson really AI or is it 'simply' a very fast, powerful and efficient sorter and analyser of immense volumes of information? Is Watson really learning?

Well, Watson had 'memorised' four terabytes worth of content, about 200 million pages of information which included Wikipedia. To help determine answers, Watson was equipped with 2,880 POWER7 processor cores (that's 90 IBM Power 750 servers x 3.5 GHz POWER7 eight-core processors x four threads/core) with 16 terabytes of RAM. Key to utilising all this hardware was IBM's DeepQA software and Apache 'unstructured information management software' (UIMA) which is a natural language processing and content analysing tool. Apparently, Watson could process 500 gigabytes of information per second which equates to about a million books per second – and he won?

In 2013, IBM uploaded Watson to the cloud. You can now interact with Watson who is no longer playing games and who has 'been to medical school'. Check the IBM website (see www.ibm.com) and note: "Today healthcare professionals are already using a diverse set of cognitive solutions that enhance and scale human expertise. Distinguished leaders in healthcare already working with IBM Watson include Memorial Sloan Kettering Cancer Centre, the Mayo Clinic, New York Genome Center, Cleveland Clinic, and The University of Texas MD Anderson Cancer Center."

Clearly powerful but amazingly, Watson doesn't rate in the world's top 500 supercomputers (in 2015). Poor old Watson was estimated to operate at a mere 80 TFLOPS (when he won Jeopardy) whilst the world's fastest computer, the Chinese NUDT 'Tianhe-2A' has an impressive theoretical peak performance of 54,902 PFLOPS[13].

To get away from having to pre determine every computation in the form of software driven algorithmic instructions that have to be written so that a computer can crunch away at high speed, scientist have turned to ANN. Why ANN - is explained by Clarence Tan (in Artificial Neural Networks): "Existing computers process information in a serial fashion while ANNs process information in parallel. This is why even though a human brain neuron transfers information in the milliseconds (10^{-3}) range while current computer logic gates operate in the nanosecond (10^{-9}) range, about a million times faster, a human brain can still process a pattern recognition task much faster and more efficiently than the fastest currently available computer."

AI however is getting stronger at an exponential rate. Computers are being taught to learn; machine learning and deep learning are terms I will explain later. An example of continuing AI strengthening is Google's AI (a computer programme called 'AlphaGo') mastering the worlds most complicated board game 'Go'.

AlphaGo fundamentally taught itself how to win and in January 2016 defeated Fan Hui, the Go champion of Europe five – nil. This is a significant step towards the achievement of machine intelligence outperforming the human mind; it is however only mimicking abilities in a narrow, all be it spectacular, field. Nevertheless, this particular feat, winning at Go, has been achieved a decade before it was predicted to be possible.

AlphaGo is however not infallible. In March 2016, in another five game marathon, this time with the World Go grandmaster, Korean Mr. Lee Se-Dol, lost the first three games, beat AlphaGo in a 5 hour showdown in the fourth and lost to AlphaGo in the final game. Clearly this represents a huge effort and investment by Google and its DeepMind project team and an absolutely commendable, lone but exceptional, human; a great media event and an example of impressive machine learning – but not true AI.

The Brain - Computer Comparison

We are increasingly exposed to fantastic computer performance statistics and comparisons with the capacity of the human mind. We are told that the human brain is theoretically capable of something like 100 trillion calculation per second and that computers will soon exceed that capacity.

Theoretical processing speed isn't everything. How fast is the human brain? I found it very difficult to find an authoritative estimation of this fundamental question until I read Chris F. Westbury's[14] post on the internet where he provided a detailed answer to the question posed by a high school student. I quote part of his response on how fast the brain processes information:

"…The best answer for this question can be obtained because we have good estimates for the three main variables that enter into it: how many neurons (brain cells) we have, how fast a neuron can fire, and how many cells it connects to. A human being has about 100 billion brain cells." [*Note that this has since been counted and unfortunately the latest estimate is that we only have 86 billion neurons – pity*].

So Westbury multiplies 100 billion neurons x 200 firings per second x 1000 connections per firing and gets 20 million billion calculations per second." [*Again, this should be 86x200x1000 = 17.2 million billion or 17.2 quadrillion calculations per second; still pretty good – I think!*]. Another property to consider is the speed of transmission, the neural operating speed; it's about 100 metres per second in the human brain but a computer operates at the speed of light, that's 299,792,458 metres per second which is nearly 3 million times faster!

Whilst Chris Westbury acknowledges that the above estimate could be "…off by an order of magnitude - that is, it might be 10 times too high or low…" he also qualifies this estimate as representing the "…raw 'clock speed' of the brain, which is much higher than the number of real useful calculations we do in a second."

Potential AI is not just dependent on processing speed but also on what this computational capacity is working on; how it is used. For that, very complex software is required and that's the challenge. Ray Kurzweil[15] has predicted that this would be achieved in the early 2020s. In introducing Martine Rothblatt's book 'Virtually Human', Ray Kurzweil explains that "The software for human-level intelligence will take longer but we are also making exponential gains in modelling and recreating the powers of the neocortex". He goes on to predict that this will be achieved by 2029.

We need to be cautious when defining the speed of the human brain and there is much debate on exactly what that is and how to better measure human brain performance. Whilst we can generalise about equivalent 'processing speeds', during complex thoughts or deep consideration, we cannot 'gloss over' the difference between raw computational speeds of machines and the exquisite intelligence mechanisms of the human brain. Here are some significant factors:

- Computers do, or soon will, process information much faster than a human brain because they have huge memory (theoretically unlimited) and fast processing capacities (many more transistors, fast switching i.e. trillions of FLOPS).

- Computers work when carrying out specific tasks like numerical calculations and are basically serial, almost linearly progressive. Although parallelism is utilised it is nowhere near the simultaneous level of the brain.

- Most of the computer's data is passive, compartmentalised, address based and non-adaptive (can be added to but doesn't self-improve without direction).

- Computers deal with facts and are fault intolerant. Even the 'Watson' experience, above, although significantly able to make sense of puns, nuance, etc. was still largely a feat of text search algorithms and logic subroutines.

- The brain, by comparison has limited memory (brain size cannot yet be increased) and fewer neurons (transistors) operating almost a million times slower (in the millisecond range mentioned above) than the, greater than nanosecond speed range, of computers.

- The brain operates continuously and in parallel (hugely simultaneous) allowing it to operate at a high theoretical maximum and is super adaptive (learning in parallel and 24/7).

- The brain has 'will' and can generalise, imagine, utilise intuition, compensate for both internal and external faults and errors, correct its own logic; it can think 'outside the box' – sorry computers – you can't!

- One of the most perplexing attributes of the brain is its cognitive abilities. From complex image recognition to the ability of instantly and correctly identify almost illegible writing, the brain's ability in pattern recognition is perhaps one of its greatest abilities.

- The brain is a learning expert whereas the computer needs to be taught how to learn. So far, the most promising approach appears to be copying and mirroring the remarkable complexity of the human brain cell, the neuron, and its neural network; and that biological mimicry may take ten to twenty years to accomplish.

These factors were correct at the time of writing but I know that many of the above are already being challenged as the development of AI and supporting, converging other technologies, are making exponential leaps forward.

V1 needs to Evolve

Back to our V1 brain: We've acknowledged its enviable capacities and why it is 'the model to be copied' in our endeavour to create artificially intelligent machines. Our brain however, just like our bodies was designed for a simpler environment. Even with its capacity and ability to learn and adapt to changing environments and increasingly complex conditions (evidenced by global evolutionary developments through the ages) there are limits to our ability to comfortably make sense of it all.

There are ample indications that our 'linear' minds (in the context of an exponentially changing technological world – see Chapter 4) are already suffering under the strain of technology, the rapidly changing environment and the sheer volume of ever more complex information to digest, to make sense of and to somehow apply to increasingly more difficult problems. Life, for the vast majority of people in the developed world, has become very complex and difficult.

We are actually 'brain limited' in our capacity to keep up with the complexities that the application of technology is bringing about. This is of course logical and to be expected. It has taken billions of years to evolve our minds. The V1 mind designed to live short but productive lives in an environment where, as Peter Diamandis says, you only knew or were concerned about what was happening within a day's walk from you.

Consider our earlier human, the V1 man from 200 to 250 thousand years ago. Even if he ran or travelled a lot, his environment was fairly limited and hence his concern and knowledge was also much localised. Same body, same sized and equally capable brain as we have today. How big was his world? What did his brain have to understand and what was he thinking?

Then, our minds were very capable of dealing with the apparently unchanging and limited environment that we were made for. The human intellect, its natural curiosity and its ability to innovate, create and radically alter the environment, over just the last few thousand years, has led to a situation where the brain, our original software, is being stretched and used for purposes that it was never designed for.

Unfortunately, now that the technology horse has left the stable, we have little choice but to cope as best as we can. Later I will deal with the transcendence from natural biological evolution to human designed and engineered technological evolution. For now, let's have a brief look at our evolutionary heritage that gave rise to our V1 human.

There is an evolutionary imperative that we organic beings cannot escape. From the time that we believe the first simple biological organisms emerged on our planet, say 4 billion years ago, about 600 million years after earth came into being, we have established that biological organisms have been on an ever more rapid journey towards complexity and diversification.

Life on Earth continues to evolve. Change, a function of the biological imperative, has on occasions been drastically accelerated by cataclysmic events such as meteor impacts and global climate changes. These events have interrupted and 'reset' evolutionary developments. The immediacy of their impacts has often left 'no time to adapt'. There is ample evidence where this 'no time to adapt' has led to mass extinction of existing and dominating life forms on the planet. Regrettably, nature doesn't seem to have a 'sorry' for those life forms that fail to adapt and it certainly cares little about what caused the change.

One notable event, the great oxygenation event which, about 2.4 billion years ago led to the elimination of the previously dominant anaerobes

31

(bacteria living in a reducing – no oxygen environment) and led to the creation of a new life form, one that no longer 'fed' on inorganic matter, one that relied on organic material (other life) for its food. That took about one billion years.

A mere 590 million years ago, these single celled eukaryotes (contain a nucleus and other organelles within its cell membrane) gave up their single existence and joined with other cells to create the first multicellular organism, the metazoan. I wonder whether there were eukaryotes that were perfectly happy with their lives and didn't want to change; why should they and why did they? Joining cells meant a life of cooperation, you look after locomotion, I'll find food and grade chemicals – I mean, who decided and why; what drove this 'beneficial for progress' development?

The result, as they say, is history. Life kept on getting more complex and with each quantum development from simple animals, bilaterians, fish, plants on land, amphibians and so on, all happening at an ever increasing rate, genus homo finally arrived about 2.5 million years ago. It was only 200 to 250 thousand years ago the anatomically modern man (the V1 model described earlier) arrived complete with a neocortex that allowed him to become the dominant life form.

Today, that biological imperative to evolve still exists. Inorganic elements and compounds were brought to life (as we can now do in our laboratories) billions of years ago. This transition from chemical evolution enabled a great biological era that developed the current abundant and richly diverse life forms on earth. Today, we are perhaps entering a new transitional era. A time when organic life will merge with the inorganic, where artificial machines, initially man made, will merge with humans.

Anticipating howls of objections to the notion that technological evolution will replace the biological evolution of humans (acknowledging that biological evolution will persist independent of this) let me just make this point: It is an almost arrogant human position to assume that as we are comfortable in our current form, we needn't change even though we are changing our environment and consuming our earthly resources at will. Our often demonstrated assumption that we are a 'final product', the V1 human model, as good as life on earth can get, that evolution is so impressed with mankind that no more changes will occur, is a self-preserving delusion and flies in the face of our evolved history, a history which has seen so much 'life form changing' that we ought to take note.

Toward V2 Bodies and Minds

That our bodies are overdue for a much needed upgrade is self-evident. Our common health issues require more and more radical intervention and interference to maintain the longevity that has been achieved through technological and scientific progress. Our V1 bodies now rely on such constant interventions and it's a circular pattern: more appropriate intervention leads to longer life requiring more care and intervention. To break free from the constant care and maintenance cycle some persistent improvements are required.

Human beings cannot stop the advance of exponential technology any more than the anaerobe could stop the oxygenation of the earth's atmosphere or single celled organisms stop the creation of ever more complex life forms. Limiting restrictions against research, laws prohibiting this or that development will not work. If it is possible it will happen; how we lead and manage the onslaught of high technology will determine our human future.

To sum up, a technologically induced upgrading of our current makeup whilst perhaps radical is inevitable by simply acknowledging what is already happening. Clearly we have stretched our ability to live our lives to the limit. We expose our bodies to extreme physical and mental performance pressures, we consume foods that our digestive system was not designed for, we expose our bodies to diseases and pollutants that we again were never designed to cope with. Although we are pushing the limits of our organic V1 makeup we nevertheless expect and seek longevity and even immortality and we are increasingly relying on technology and specifically machines to get us there. Ask someone with an artificial heart whether humans enhanced with machines is a good thing. Similarly, hearing aids, eye glasses/lenses, prosthetic limbs are artificial enhancers of our human condition. In fact we have many humans today who are already V1.1 or V1.2 humans.

It is not surprising then that we will absolutely pursue whatever technology will potentially enhance our survivability and longevity. This will involve radical developments and treatments including:

- Genetic modification that will repair, renew or regrow failing organs and tissues
- Additive manufacturing of body parts and organs (e.g. replacement - 3D printing of kidneys, hearts etc. and enhancements like additional organs that may be mind-enhancing or additional biological processors like energy cells processing inorganic materials)
- Upgrading of physical functionality (physical performance enhancement e.g. on demand oxygen/energy, artificial skeletal and muscle components)
- Upgrade of mental capacity (e.g. cloud based access to extended neocortex) on an as required or permanent basis
- Human machine integration which will mark the advent of the 'neobeing' and characterised by permanent integration of nano-organisms working alongside organs and providing real time conscious performance feedback on body state and organ condition)

Feel free to add to this list – if you can conceive it as an apparently beneficial human application it will be created or catered for.

The take away here is that it is a fundamental characteristic of life on earth to adapt and evolve. Technological advances are not accidents, they are the product of man's irresistible curiosity, intellect and desire to solve problems; inherent traits that will continue to create complexity and diversification.

The above 'natural' progression towards a 'V2 neo-human' might represent a likely future but there are some key global factors and impacts that might accelerate or completely stop our evolutionary process. Our global environment and governance may let it happen in a managed and controlled manner and either share the benefits, monopolise the benefits, resist the technological development or indeed bring it to a complete halt. We'll see soon enough.

CHAPTER 3

Human Clustering, Values and Subordination

In Chapter 1 I wrote about the necessity of considering 'being human' in order to answer the 'so what' and 'why it matters' when thinking about or shaping our future. In the previous chapter we looked at our physical makeup and the continuation of evolutionary progress. None of this of course happens in isolation. How we as humans interact, what we value, our individual and social needs, our cooperation and pursuit of human enterprises are key influencers in how we as a global community govern ourselves and our environment.

Yes, yes - we all know the origins and the creation of increasingly more complex societies, nations etc. and we all understand social interactions. Perhaps, but most of us are not consciously aware of the sacrifices that we make, the subordination of self-purposes in order to fit-in and conform to societal norms. In this chapter I want to bring this into our neocortex and show these factors to be 'the cost of belonging' and by default the cause of difference and conflict. Humor me in my simple technique.

The Lone Human

Imagine the lone hunter-gatherer; say more than 150,000 years ago. Clearly, not likely to be really 'alone': Where are his parents? What, he just materialised? Just imagine that he was in fact alone. Focused on survival, finding food, a mate and constantly on the look-out for danger; that hunter is in total agreement with – well, himself and his immediate environment.

Sure of his purpose, he feels justified and certain of the correctness of his decisions, his actions and their consequences. He is happy to learn and observe; he has challenges but they are all his and his alone. He's happy and capable of living his self-determined purpose: to simply meet his daily needs and to survive to the next sunrise.

Even back then though, our hunter was mindful. He already has a well-developed brain (neo-cortex) and is able to imagine, plan and perceive possibilities. He is a thinker. He's putting his thinking skills to good use. He hunts well; he anticipates events and continually outwits his prey. He is insatiably curious, a keen observer of all he perceives and he's rationalising everything he experiences. The world is an exciting and mysterious place to him and he uses his reasoning and imagination to fill in the gaps between what he knows for sure and what he cannot yet explain. He only concerns himself with things that are within a day's walk from him. He doesn't know what's happening anywhere that is beyond his horizons – but he thinks about it.

Whilst his mind is flooded with questions, he continues to define his reality in terms that make sense to him and he is not in conflict; he has his objective reality based knowledge; he doesn't question the mountains, the rivers, the animals, the vegetation or anything else he perceives to be real. Where he has thoughts about things he doesn't understand he forms what to him are plausible beliefs and explanations; beliefs that support his observations, that he considers against his learnt wisdom and objective experiences. Not surprisingly he is in total acceptance and agreement with his formulated beliefs.

Our loner lives immersed and in harmony with his objective reality. An objective reality that is however harsh and full of peril; he is necessarily forced to face survival challenges around the clock. His imaginings don't significantly affect what he does and how he lives his life. He knows his purpose; no ifs or buts. He is accepting and happy in his clarity of purpose!

Two Humans

Now there are two. Two individuals that have joined in some bond and for some perceived and shared purpose. The human social experiment begins.

Let's set the clock to say 100,000 years ago. Our two humans have language, symbols and a developed appreciation of their environment and shared beliefs about their reality. They are self-ware with a sense of identity and purpose. These two individuals have joined for very real and objective purposes. 'Safety in numbers' starts here. Mutual support that greatly enhance their abilities to survive in their physical environment, mating, cooperation, the joining of skills, experience and knowledge; many clear benefits for staying together and cooperating.

Beyond the obvious survival benefits of this group of two, their teaming up facilitates the first sharing of ideas. The sharing of perceptions, experiences, knowledge through mutual communication heralds the era of information sharing. Ideas can now be shared, evaluated and built upon; ideas that enable the realisation of higher values as their mutual support has freed up considerable time for thinking, planning, scheming and experimenting. The ability to ideate, to experiment, leads to greater innovation and evolutionary progress is accelerated.

I don't think I need to spell this out; you get the picture. 'Two minds are better than one' is not a controversial concept. So what then?

Firstly, it is clear that these two are influencing each other. Clearly when two people agree to cooperate, to team up, to join forces one of the two leads the other. Please don't believe that they are instantly self-organising even if instinctive reactions and behaviour might indicate this. For example, a sudden threat appears, one ducks or cowers and the other faces the threat aggressively. Self-organising or reflecting a realistic learnt survival response (the reptilian brain: flight, fight or freeze response)?

One will influence the other to follow. They will accept that whatever their objective might be, one of them will be better suited to take the lead. The other will agree and subordinate their own skills, opinions and desires if the other's leadership is assessed as being the most likely to successfully meet their joint objective.

This needn't be a permanent arrangement. Their next joint objective might require the previous follower to take the lead. In circumstances where one of the two has a persistent advantage, leadership of one over the other will become permanent and dominating.

Whatever the leadership situation might be there is one truth to be recognised. These two each have their own and unique value systems. These values, however firm or strongly held, defined or vague, may be similar or diverse but understanding that no two minds are identical, they will be different. They do have their own beliefs, thoughts and evaluations; their very individual experience based existential values system. To the extent that they are aware that these direct their thoughts, actions and experiences, they adopt a sense of individual purpose. To a lesser or greater extent they define their own, personal 'Telos'

This is important to understand so let me detail what I mean here in the language and wisdom of the 21st century:

Telos - Higher Purpose

Telos[16] (from the Greek τέλος), is defined as the purpose, end, aim, or goal of something. Dr John Demartini[17], in his book, 'The Values Factor', eloquently uses this term to represent one's very highest value; one's evolving purpose here on earth. He explains that it is this that "inspires you most – the purpose you feel grateful for simply pursuing. Your truest and wisest way of serving the world is to do the thing you love most. That is your spiritual path – and your inspired destiny."

So 'telos' is your greatest purpose, the most important thing in your life; it is your very unique mission and reason for being. Influence (lead) yourself to discover your highest value.

We should all live according to our own higher values. Unfortunately, largely due to unwarranted comparisons, too many live according to other's values. Noting how successful someone is might lead you to adopt their value system; a spouse might be dominated to live according to the values of their partner and totally subordinate their values. Most live a life that is constrained by social conformities, social idealism, cultural traditions and unchallenged paradigms.

You cannot be a leader, an influencer or even a worthwhile individual, if you subordinate your purpose to others. To be an effective leader you must have a degree of mastery over your life and this can only occur when you have discovered you higher purpose. Your ability to achieve is directly linked to the level of alignment with your telos.

How do you know your higher purpose? Examine what you enjoy, what inspires you, what do you love to do? Consider what you spend time thinking about and what is the subject of any self-dialogue? Look at your life – it demonstrates your values. Having discovered your present telos, acknowledge that it might change over time, it evolves; give yourself permission to be the unique you that you were created to be.

Understanding your values will drastically affect your life. It should severely influence your thoughts which will direct what you sense, focus on, spend time doing and shape your destiny. Your brain will interpret your experiences according to your higher values and your thoughts will direct your actions that will realise your telos.

Leaders need to dwell and operate primarily in the neocortex. Living with an awareness of your higher purpose means just that. To simply follow your passion may lead you to an unhealthy engagement with your reptilian mind that is animalistic, ungoverned, emotional and preoccupied with self-gratification. Whilst the reptilian brain remains vital to survival it is not where your thoughts should dwell.

'Follow your passion' is never good advice if unqualified. Your passion could be animalistic, anti-social or destructive. Thoughtful advice would be to align your mind and your passion; in other words – engage your neocortex, think and be 'mindful'. When you decide and define your higher value, your higher purpose, you will allow your mind to filter all experience against those values; it will bring to your attention factors, people and actions that will guide you to achieve your telos. Dr John Demartini says "your brain is a telos fulfilling organ".

Being aware and living your higher value will dramatically affect your feelings and your thinking, so much so that it will restrict the influence of emotion and physiology. It is difficult to unsettle someone living their telos. Living your telos will also define your associations, friends and people with whom you will have personal relationships. You will naturally and mutually be drawn to people whose values are closely aligned with yours.

Defining your telos defines you and shapes your future. Finding your purpose, defining what makes your life worthwhile for you, is critical; don't leave this fundamental driver to chance or default.

We all have our own and very unique value systems. When I refer to your higher purpose, your telos I'm referring to a deliberate and considered purpose; one that you've adopted for your own purposes, desires and passion. Many of us however haven't done that. We might say we have no self-aware purpose, aren't particularly passionate or mindful about anything or indeed are passionate about lots of things. That's OK but nevertheless we all, consciously acknowledged or not, have a value system. We all know what we like and don't like. We all feel that we do or don't like certain people, activities, situations, environments and so on. These default values are the result of inherited values and experiential beliefs; these are our values.

My use of the term 'higher purpose' includes these 'values'. The John Dimartini definition above is a measure of the quality, the life motivational energy, the fulfilment, drive and joy that a realisation of one's unique value, the higher purpose or telos, will bring on realisation and adoption.

The other very important point to acknowledge is that of labelling. Labelling is the external designation of a person into a certain category. This usually happens to children as soon as any comparative situation arises. Johnny is a thinker he's going to be a doctor one day. Maria is a doer; she's always making things; she'll be a tradesperson (or worse will make a really good housewife one day!). Too many people accept such labelling to some extent and such acceptance, when continually reinforced, becomes the conditioned, the default value system that influences the majority of life's decision made from then on.

In the 'tribe' section below I'm going to have the tribe assign a warrior role to our previously lone hunter. He will accept the role and probably live his life conforming to his labelled role – the warrior. Would he be fulfilled in that role? Would the inner desire to act on his passion for hunting be realised? Would he realise his potential and consider his life all that it could have been when he draws his final breath? What do you think? Why?

Back to our team of two: Developed individual telos or not, they each have, to some extent, made a compromise; a willing subordination to the other in forming a team. Accepting this, however beneficial to the group, is the first step in the very logical establishment of 'difference'. It represents, to some extent, the forgoing of personal desires, beliefs, objectives, actions and behavioural preferences and individual freedom for the good of the group. For our two humans, that's not perceived as a serious sacrifice because the benefits so clearly outweigh anything that is consciously given up or subordinated.

Mathematically, let's express this as $HP^G = f(dHP^{I1}/t + dHP^{I2}/t)$ Where HP^G is the group higher purpose expressed as a function of the higher purposes of the individuals in the group, HP^{I1} and HP^{I2} are the individual's higher purposes, t is the duration of the agreed modification of the individuals' HP and the differential is the measure of subordination

I'm expressing this mathematically because I will explain below the degree of compromise that any one human is potentially subjected to, and to varying degrees forced to accept, when the number of individuals is 7.4 billion.

The key here is to note that the individual's higher purpose has been perhaps altered or at least supressed, and to some degree, prevented its expression so as to conform to the group. Willingly or not, the individual's higher purpose has been influenced by the need to accept the group's purpose in order to be and stay a member of the group.

The Tribe

Accepting that our team of two, humanity's smallest grouping, was a net beneficial arrangement; required some perhaps small and apparently unimportant giving up or subordination of individual values it also introduces the seed of disagreement, conflict and established points of difference.

Let's consider the higher purpose, the values factor in a larger group. A group like a multi-generational family grouping or a tribe of several such families. We can accept that each individual of this tribe has accepted and to a greater extent adopted the group's higher purpose. After all, that is the dominating reason for being a tribe. We would accept that the degree of subordination of any individual's beliefs and values would be commensurate to their standing in whatever organisational, governing and leadership structure had been adopted by the tribe. The strong leader has most likely not significantly altered his or her higher purpose. Simply and obviously - their ability to lead the group, their ability to influence the group is why he or she is the leader and it stands to reason that the leader's beliefs, intentions and purpose for the group will predominate.

At the other end of the scale, a weak tribal member will not be heard and their higher purpose will be subordinated without compromise. My use of the term 'weak' represents a value judgement that already evaluates that particular member's potential to support and contribute to the tribe's purpose; in assisting the tribe to achieving their objectives.

The advantages of the tribal group cooperative are significant. Regardless of the leadership style adopted, the tribe will assume a workable organisation and management structure. The structure may not mirror the 'plan, organise, control and lead' model of our fading industrial age but it will nevertheless direct the operation, the functioning, of the tribe in a cooperative fashion. The tribe will in fact be marshalling and allocating its resources to realise its objectives; to live its tribal higher purpose.

The lone human had to fend for himself. He had to meet all his survival needs alone; a need that allowed little time or energy to engage in higher level thinking and experimentation. The team of two fared better; sharing of survival related tasks made the job easier for both and they could also share ideas; the prerequisite to accelerated innovation. The tribe is of course much better at meeting its collective survival needs than our loner or team of two.

Physically, the tribe develops specialisations where those most able, appropriate or directed to carry out specific tasks get the job. This is the onset of the diversification of skills, skills that will establish weighted values to the goods and services provided to the developing tribe. These weighted values will reflect the specific challenges of the tribe. A tribe that is facing a food shortage will place high value on their hunters, gatherers, farmers etc. A tribe threatened by a possible attack will obviously value and willingly sustain its warriors. A tribe that gets its skills diversification and allocation correct and allocates sustainable resources accordingly will greatly enhance its chance of survival.

Having eased the requirements to exclusively focus on the tribe's survival needs, individual and collective attention will increasingly turn to higher value pursuits. Again, let me leave the tribe for a moment to explain this.

Human Needs

Human values logically reflect human needs. It's clear and obvious that we need to have what we actually 'need' to survive before we need anything else. The link between needs, human values and motivation has long been known. The most recognised theory on human needs and their relative ranking was presented by the American psychologist Abraham Maslow in his 1943 paper titled 'A Theory of Human Motivation'.

His theory represented as a pyramid, titled 'Hierarchy of Needs' ranked categories of innate needs. He argued that these needs are required to be met 'from the bottom up' in order to achieve self-actualisation. They are (from the bottom up): physiological needs, safety and security needs, love and belongingness needs, self-esteem needs and self-actualisation. It is this self-actualisation that I have previously described as 'Telos'. Living your potential, meeting your self-defined higher purpose is your ultimate self-actualisation.

More recently, Peter Diamandis, in his book 'Abundance', re-interprets Maslow's pyramid and absolutely nails the meaning of Maslow's ultimate needs, the need for self-actualisation, when he writes that they "are about personal growth and fulfilment – though they really constitute one's *devotion to a higher purpose and a willingness to serve society* [my stress]."

He defines his own three layered 'Abundance Pyramid' in terms of progressive requirements to be met. The three layers of his pyramid (again from the bottom up) are:

Physiological needs (food, water, shelter)

The pursuit of Catallaxy (catalyst for further growth; energy, education, communication and information), and

Health and freedom (factors that strengthen the individual's ability to matter and hence contribute to society)

Meeting these requirements will enable progress towards the realisation of abundance; the core message of his book: that abundance for all is possible and that abundance means "...providing everyone with a life of possibility". A particular strength of his work is that he stresses the inclusiveness of abundance and that "the individual must matter and matter like never before". It is also important to note that the Diamandis 'peak of the pyramid' is not defining a self-centred state, rather it puts personal actualisation into a socially beneficial, a positive and contributory, context – "to contribute to society".

The take away here is that both appreciations of the hierarchical needs of the individual and of society point to a higher purpose; a purpose that is both self-actualising and adding value to society. This is the mindful alignment of personal higher purpose (the individual telos) to that of the greater, social or group values.

The tribe now turns its attention to higher level needs. Adopting the Diamandis model, these are the needs for energy, education, communication and information. Meeting these higher level needs will enable the tribe's individuals' potential to grow, contribute to the advancement of the tribe and reinforce the tribe's ability to achieve its adopted objectives – its higher purpose. Note here that a higher purpose is something that changes with time and reflects current thinking, wisdom, learning and the appreciation of possibilities. This is true for both individual and group telos.

So what's happening to the individual's higher purpose? The tribe has now greatly diversified the roles of its members. It has created a value system, choice, opportunities, culture and ideologies. The tribe's appreciation and accepted understanding of its physical reality has been defined. All members of the tribe have accepted and identified with the tribe's telos, the tribes higher purpose – or have they?

The advantages of belonging to the tribe are compelling. Given that the case for belonging is so self-evident, most Individuals willingly identify with and support the tribe's objectives. Some of the tribal members however might have some reservations. For example, our lone hunter might not have been selected to be a hunter; he might have been persuaded to be a warrior because the group valued his fighting skills above his ability to hunt and fish. Although our loner, now a tribal warrior, holds a valued position in the tribe he nevertheless harbours a degree of unhappiness and mild resentment to his allocated job. He misses hunting and the freedom he used to enjoy. He feels uncomfortable with the need to always do what the tribe decides.

His personal higher purpose hadn't changed. He has somewhat reluctantly subordinated his own goals so that he could be a member of the tribe. Our loner is however not unique. All members of the tribe have had to subordinate some or all of their higher values to that of the tribe. Giving up on personal desires, preferences and higher values is the cost of joining the tribe for the common good. As long as the common good remains just that – good for all of its members, then that sacrificial subordination of individual wants is accepted, even if somewhat begrudgingly.

Tribes Meet

Our original tribe has fared well. Having successfully met their survival needs, the tribe is growing rapidly in number, continuing to develop its technologies, education and culture. The tribe enjoys a balanced existence. The members of the tribe have assessed their reality with an insatiable curiosity and a desire to find meaning in all they observe. They have recognised the day night cycle and have come to appreciate the warmth and light that the sun is blessing them with. They've agreed that they are indeed the children of the sun. Their symbols of their sun god are everywhere and their belief is reinforced by their leaders who will tolerate no disrespect or challenge to their tribal belief system.

The religion of the sun is now a dominant characteristic of the tribe's telos. All tribal activities need to be in accord with the perceived dictates of the sun god. The tribe's leadership has accepted their divine responsibility to bend the tribe's higher purpose to that of the sun. The tribe's leaders work closely with their religious leaders; they understand that many of their tribe are very religious and that they need to reflect their staunch appreciation of the sun in order to remain in leadership roles. I'm sure you get the wider implications of this.

Our 'sun tribe' expands its area of influence, foraging and exploring increasingly remote locations until one special day. The scouting party of the sun tribe discovers strange totems in the jungle. They recognise these symbols. They have encountered the 'tribe of the moon'.

Clearly, you can picture the likely continuation of the story. Many probable outcomes are possible: all-out war, the destruction of either or both tribes, long periods of hostility, avoidance and retreat etc.

Our history proves that similar meetings of civilisations took place and that at least in some instances an eventual joining of tribes took place.

In our sun and moon tribe example, obvious difference would have been apparent to all: differences in skin colour, body shape, dress, food, shelters, tools (technology), language, organisation, behaviours, culture, and values and so on. These differences, once appreciated as fundamentally enriching to the new civilisation, were embraced and built upon. The opposing sun and moon religions were demoted and replaced with a new and accommodating belief system. Both previous religions were deemed to have been valid and became the common justification for the evolved 'religion of the sky'.

The citizens of our civilisation prospered. They lived in this new super tribe assured under its governance regime that their full range of needs would be met. And their needs were now very complex and certainly no longer solely focused on the bottom of the pyramid. This civilisation continued to evolve and eventually had created a complex web of conventions, systems of rules that covered all human interactions, activities, ethics and beliefs.

The citizen became totally caught up in these systems. Their reality had changed almost without notice. Their desire to realise increasingly loftier goals and to free themselves from the burden of survival needs led them to eagerly adopt these created systems.

What then are these systems?

Objective and Artificial Realities

We all live in an environment that constitutes and defines our reality. Let's consider this reality in simple terms. A suitable definition might be that our objective reality constitutes all that is perceived by everyone; our existential physical world and everything in; the landscapes, the vegetation, waterways, lifeforms etc. – all those real features that make up our natural environment and directly impact our ability to meet our essential needs, the bottom of the needs pyramids and the critical requirements for our survival.

In simple, primitive or 'underdeveloped' societies (like the above two human and one tribe examples), their reality is fundamentally objective. But with evolution, the development of the powerful neocortex which enabled advanced cognition (the earlier mentioned abilities for planning, modelling, simulation, reflection, language, abstract thinking, and perception) we humans developed another reality. This reality is non – objective; it is the imaginative reality, an often variably shared fantasy that consists of human constructs. This is the reality that only humans perceive. This is the fantasy of politics, conventions, culture, religion, ethics, money, rights and obligations, social norms and expectations.

A simple test to decide whether any human construct is real or imaginary: pretend to be trading with a chimpanzee. A chimpanzee has a neocortex and can think, albeit not at human levels; nevertheless the chimp, let's call him Wally, is smart enough for this test. Does Wally accept money as real? Would he swap a banana for a dollar note? Would Wally give up something of value, say a bag of peanuts, for a promised salvation after his death? Would this chimp be impressed with your copy of the Universal Declaration of Human Rights? Of course not! Wally would probably reject all we offered with words like: "Don't be silly; what do you think I am; human?"[18].

I'm not for a moment judging these 'imagined' systems, I'm just defining the difference between what is an objective reality (the reality perceived to exist by all advanced species) and what is a very human, imaginary, created reality (only perceived and valued by humans – human constructs). This is critical to understanding the world today. None of us can deny our physical reality, certainly not Wally the chimp; but our world has become much more complex than that. Most of us have become so far removed from our common reality that modern life is intolerably complex; full of conflict, confusion and apparently, increasingly irreconcilable differences.

We should agree with the observation that our life experience is necessarily founded on our objective reality – the 'real world' our physical world. We need to also understand that at the same time, we all live in an imaginary world; a world that only makes sense to us humans; this imaginary human world of our own construction might be termed our artificial reality.

So what then? What is the connection between human value systems like the pyramids of needs, higher purpose and these two realities?

Let's consider our joined tribes' civilisation above. We have established that the individual's higher purpose, their telos, has been to some extent subordinated to that of the organisation. We note that this is not openly resisted by the individual. The civilisation bestows advantages to its citizens and meets their needs. We acknowledge that the increased resources, progress and developments of this civilisation provide its citizens with not just bottom of the pyramid needs but needs that lead to the apex – higher needs leading to self-actualisation.

It is however critical to understand that the individual's higher purpose has necessarily been subordinated in every case where the greater group's (be it tribe, civilisation or nation) higher purpose is not directly in accordance with that of the individual's telos.

Given the very individual and unique higher purpose of every citizen it should be clear that the civilisation's higher purpose will be increasingly difficult to define, it will be divergent and distant from those of its growing population. Why?

'Higher purpose' reflects the environment. In a simple society, one firmly anchored in an objective reality, the higher purpose of individuals and the society will not be vastly different. It is clear that values and higher purpose will be focused on the survival level and anything above that will nevertheless be directly and perceptively linked to that objective survival environment. This is evident in considering any of the very 'underdeveloped' nations or remote tribes that exist in our world today. We all agree that they live 'close to earth', in harmony with nature; we don't always acknowledge that they are living in their objective reality.

The Global Tribe

In early evolutionary groups, such as the above tribal groups, resources were easily shared, equality was achieved within the limits of basic hierarchies, all were identifiably 'in the same boat' (shared very similar life defining circumstances), and the fate of the individual directly affected the functioning of the group. There was a very close and direct link between the individual's value and the value of the group. In these groups, mutual care and compassion was critical. The group did not leave the individual behind because the individual mattered. Children, a mate …all were important to the group and usually demanded significant investment. Teaching the young for example was of key importance. This ensured the long term survival of the group, the preservation of tacit knowledge and culture. The wisest, strongest, the best of the group were the mentors, tutors and teachers of these groups.

A key attribute of early evolutionary groupings was shared understanding. All members understood their environment and could survive by sustaining a livelihood. They had the knowledge and skills to meet their survival needs.

As tribal groups grew specialised skill and function based sub-groups developed within the tribe. Each sub-group had its own leadership, defined purpose and members that subscribed to the sub-group's purpose within the overall purpose of the greater tribe. Relative sub-group advantages and disadvantages would wax and wane reflecting progress, technology, changes to the environment and interactions within and external to the group; in short, groupings became dynamically adaptive to existing paradigms and needs.

If we acknowledge that a typical nation is one that comprises many diverse tribes, communities if you like. We have individuals, families and groups of people living together, societies, suburbs, villages, cities, counties or municipal groupings, states – a variety of constituents that might make up a nation. Within and across these groupings we will find a wide range of purposes that support the existence of these groups. These include, ethnic, religious, professional, trade, political, social, lifestyle, rich, poor, gender, age related, recreational, philosophical, indigenous groups….- a long list of belief, economic, social, environmental and political groupings.

Just as our tribe was formed because its members shared a fundamental and dominant need and hence belief; our diverse groups also share some common purpose for existing. This purpose is the group's higher value and to some extent its members agree with that higher purpose and it forms part of their individual telos.

Each of our modern day groups then will be a unifying influence to all of its members. It will represent a common bond and any outside influences will be assessed against the group's interests. Now these groups of course exist within other larger groups and it is logical to expect that their belonging and conformity to this bigger group needs to be a factor in the higher group's purpose. For example, a social activity group cannot sustainably exist if its activity is not supported, tolerated or catered for within the larger group.

I made the point earlier that an individual's higher purpose usually requires some subordination in order to fit in with a group. I also pointed out that this 'fitting in' is dependent on the alignment between individual and group higher purposes. In modern society this is no different. The same dynamic applies. Replace the individual with a group and then place this group in an even bigger group and keep doing this until you've included all global groupings.

The individual's higher purpose is successively subordinated (by default, willing or unwilling) in direct proportion to the number of individuals within any group and the number of layered groups that make up the largest group of which our individual is a part.

Subordination

Let's confirm again what I mean by subordination. Fundamentally it is the acquiescence to group think; it's a compliance issue; complying with the values and norms of the group. It mostly doesn't 'feel' like a sacrifice or any kind of subordination. The benefits of belonging dominate our thinking. Most of us never examine our core beliefs and don't ever define a purpose for ourselves. We might belong to groups by virtue of geographical location (race, ethnicity, political position etc.). We might vote according to inherited beliefs (dad always voted liberal/labour so I do too), I don't like the rich because I don't have money and therefore rich is bad, I was born into the sun religious family and I know that moon worship is evil.

Simple, but please accept that most people never evaluate their beliefs and even more rarely change their views to a more objective one until it becomes an apparent threat, a survival impact. In most cases an individual's higher purpose is not a conscious state of mind; unfortunately it's the default, group influenced purpose that is adopted often for no discernible good reason. Running with the sheep is a reality. Belonging to a group and accepting group think i.e. the group's purpose is the default position in which the non-thinker draws comfort and has subordinated the reason for his or her existence. Not all black or white of course and dare I say it – many, many shades of grey.

Clearly, advanced humans (mostly those that our world would regard as successful) have decided their telos. They live their often unique dream; they have defined their values factor, their higher purpose – their telos. They choose to belong to larger groups and they subordinate only the 'must do's' and they do this in full consciousness.

The take away here is that in order to belong to a group, a degree of conformity is required. Be this conformity deliberate, begrudging, and resisted or otherwise, it represents a subordination of your individual aware or unaware purpose.

Ask anybody what they value in life. The most common response, regardless of the initial answer will usually be qualified with something like "as long I have food, health and a safe place to sleep…" I'm sure that this is your experience too. It's obvious that we are all intuitively aware of our common survival needs; our dependence on our objective reality.

Ask people what things they would want to have in a post-apocalyptic environment or on a deserted island and they should (unless they're stupid) list survival tools and materials.

Our needs reflect our specific environment. Our desires reflect our default or deliberate purpose. Our life then is dependent on both: our environment and the groupings in which we exist. The quality of our lives depends on the degree of subordination; the degree to which such subordination is seen as beneficial, chosen or imposed and the appropriateness of such subordinations in a dynamic and changing life.

Returning to the global tribe; we have over 7,400,000,000 individuals in this tribe. This is our tribe. Our only common characteristics are that we all live on earth and that our undisputed survival needs are the shared dependencies on food and water. These two facts are the only factors that all 7.4 billion of us cannot yet escape. Anything additional to these existential common needs starts the magnificent diversity and despicable conflicts that defines us today.

Earlier I expressed the higher purpose of a two person group as:

$$HP^G = f(dHP^{I1}/t + dHP^{I2}/t)$$

Where: HP^G is the group higher purpose expressed as a function of the higher purposes of the individuals in the group, HP^{I1} and HP^{I2} are the individual's higher purposes, t is the duration of the agreed modification of the individuals' HP, and the differential is the measure of subordination.

How will this equation look for the world group?

$$HP^W = f(dHP^{I1}/t + dHP^{I2}/t + dHP^{I3}/t + \ldots\ldots\ldots + dHP^{I7,399,999,998}/t + dHP^{I7,399,999,999}/t + dHP^{I7,400,000,000}/t)$$

Clearly, defining a possible world's higher purpose would need to take into account the purpose of every one of its 7.4 billion people. How possible would it be to find a common purpose other than those that provide for basic survival needs? If we now consider that the experiential reality (objective and artificial discussed above) varies significantly according to how 'developed' national groups or individuals are then how would that even make sense?

It doesn't get much better if we consider small nations. Say Australia with a population of 24,309,330 people (as at Jan 2016). The equation would be:

$$HP^A = f(dHP^{I1}/t + dHP^{I2}/t + dHP^{I3}/t + \ldots\ldots\ldots + dHP^{I24,309,328}/t + dHP^{I24,309,329}/t + dHP^{I24,309,330}/t)$$

Still over 24 million individuals all trying to define, agree on and live by a national higher purpose. That is: what does it mean to be Australian? Do they all agree and what personal values or purposes are, or need to be, subordinated in order to conform to the Australian group purpose. The HP^A, Australia's higher purpose, isn't a single variable; it's a mixture of identifying principles, religious and cultural practices, ethnic characteristics, policies and legislative frameworks, the rule of law, certain human/citizen's rights etc. However, the fact remains, to be a part of this or any group, some degree of subordination is required.

"You can't please everyone all the time" – clearly not; it's difficult enough to please everyone even in the smallest groups. We'll come back to this when we consider governance in more detail in Part 3.

CHAPTER 4

Experiential Realities

In previous chapters we've looked at human values, values that relate to our hierarchical needs which can be ranked and categorised (e.g. Maslow's or Diamandis' pyramids). We can deduce that as lower level needs are met, human focus shifts to higher level 'needs' and that values are determined by these perceived needs.

We've established some grouping mechanisms that relate to human values. Human, needs based values determine groupings and associations. It is clear that the more advanced human groupings are, the higher their focus or purpose will be. Equally, choice, opportunities, desires, wants etc. will increase and diversify with resource availability that result from having met lower level needs.

Values also determine purpose. Our individual values might be sub-conscious default values, deliberate telos or higher purposes or anything in between. It is important to acknowledge the uniqueness of purpose; individuals are unique and although individual values can be very similar and near identical they are nevertheless different.

We need to recognise that groupings of individuals requires acceptance and hence some subordination of individual values. Subordination is simply a consequence of group membership and regardless of the degree of subordination awareness it too can be deliberately accepted, accepted sub-consciously, begrudgingly and even imposed. Regardless of how voluntary or acknowledged subordination is it will impose some non-realisation or restriction on personal wants, purposes or needs.

I want to stress that there is a very simple logic to understanding that the grouping of humans will necessarily require subordination. Remember that this subordination relates to the extent to which an individual accepts limitations to 'living' their values, their telos in order to belong to the group.

Groups of course are defined by their apparent formational purpose, their reason for being. This group purpose might relate to social association, environmental condition (geopolitical, geographical etc.), religion, race; anything that to some extent unites individuals in the group. We note that a group's higher purpose will reflect some common values of individuals in the group.

Our global population today is above 7.4 billion people[19]. This population is made up of an almost endless number of groups. Name any country, religion, race, consumer demographic, generation, corporation, enterprise, professional association, school, business, club etc. and the multitude of groupings becomes apparent. Each of these groups has a defining purpose. To some degree each group has a telos and its members, again to some degree, share in those values and hence belong. These groups are however dynamic. The rise and fall of civilisations, popular culture, what's 'in' or old fashioned, rebellion and change in general are all clear examples of dynamic group systems.

We acknowledge that individuals may be members of many groups and that just as individual purpose evolves with time so do the objectives and purposes of groups. Groups then are truly diverse, dynamic and a fundamental and constantly changing organisation of association, of human sharing of values. Values that in themselves vary in response to individual needs, circumstances and desires; needs and desires that drive humanity in a reality that is both objective and artificial.

Let's look at this in an environmental context; specifically the environmental context that determines our existential experience. The environment that we individually perceive as the world we live in.

Perceived Realities

I've previously introduced the concept of the real, natural, objective, physical world – the world shared by all life on earth and one that is for the most part equally perceived as 'real'. I also introduced the imaginary world, the artificial reality. This is the human fantasy world of our constructs; something we have created. This is the world of religion, money, politics, governance, law, human rights, marriage – even lies, ethics and knowledge. Objective reality is naturally perceived and there is little disagreement about it; imagined reality, that artificial human construct, is however subject to radically different interpretations and subjective evaluations.

We perceive our world as a mixture of both the physical and the artificial reality. We assess this world according to our unique circumstance, values and purposes and further appreciate our realities through the filters applied by our groupings. This is important.

At one extreme, the reality dominated and almost totally objective (physically real), we might have primitive culture. This could be a subsistence agricultural village or a nomadic tribe. For these individuals their life experience would primarily be focused and dependent on their physical reality. They would live 'close to' and 'in harmony' with nature. Their daily existence would be dominated by landscape, vegetation, weather, and seasonal cycles; their ability to meet survival needs would directly reflect their understanding and knowledge of the flora and fauna that share their world. Their reality and hence their values and aspirations would reflect their physical world. They would however have some advanced human constructs (artificial reality) and this might be in the form of a level of spirituality, cultural or social practices - but the balance of objective and artificial reality would be in the order of 80/20 respectively.

At the other end of the spectrum, where the majority of the individual's perceived reality is in the artificial zone, we have the highly 'developed' modern human living in the most technologically advanced nations. Here, life is almost totally dominated by human constructs. Money, law and order, social conventions, education, employment, wealth, status, material goods and possessions dominate the perceived world. Modern cities appear to be totally unrelated to location, geography or climate.

Human constructs have effectively terraformed local environments to accommodate the needs of the artificially created world. This is no criticism; it's just an observation that reveals what dominates life at this end of the reality spectrum. An example of this is that a hotel room of any global hotel chain is almost identical whether it is in Los Angeles, Bahrain, London or Sydney. You can literally wake up in any of those rooms and not immediately appreciate where in the world you are. Artificial constructs have effectively created environments that are independent of physical realities.

Survival here, in the artificial reality domain, has virtually nothing to do with the real physical world and objective reality. Here, individuals' values depend on their interactions with the human constructed reality. The natural environment plays no significant role in adopted purpose and higher level needs dominate life in this 'fast lane'. Of course basic survival needs are necessary but these too depend more on the correct functioning of artificial creations (advanced collection, processing, distribution and retailing of resources) than the local objective reality.

7.4 Billion Compromises

So we might consider that each of the global 7.4 billion humans experience a different mix of objective and artificial reality. Clearly we need to acknowledge groupings here and we might consider national groupings. We might see under-developed or developing nations that are largely agricultural societies at one end and highly developed nations at the other of the reality spectrum. In each case, the national higher purpose, national values will reflect the communal values to a greater extent but there will nevertheless be great diversity among the individuals that make up such national groups. The common good: the 'good of the nation'; the national interest now becomes a very complex issue.

In the advanced nation for example there might be significant numbers of impoverished individuals and whilst their national higher purpose and values reflect the needs of the majority and be focused in delivering high value needs, the impoverished may be facing needs that are very much focused on basic survival.

For each of our 7.4 billion individuals there exists a unique profile of needs and value definition. Let's consider a profile of eighty percent objective and twenty percent artificial reality. In this 'mix' the individual has the ability to realise their own complex value system. Whilst living in this 80/20 objective/artificial reality world, the individual nevertheless has the opportunity to move through the layers of hierarchical needs and achieve total self-actualisation. The definition however of what constitutes self-actualisation will almost certainly be radically different to what might pass as worthwhile (subjective) accomplishments to an individual existing in a different reality mix environment.

In Part 3 I will return to this issue and its relevance to governance and government. The above should have established that every individual has a unique view of the world and their values; their place in their perceived reality presents unique challenges and opportunities. Individuals adopt purposes, aspirations and form concepts of a rewarding and fulfilling life based on their unique appreciation of their existential reality.

Such individual adoptions of values will to some extent conform to group higher purpose but individuals who meet their basic needs will increasingly consider their individual likes and dislikes; their differences. With a developing awareness of subordination to the group, individuals will examine their 'compliance costs' of belonging; this is the onset of divergence and conflict resulting from the contemplation of ideas and concepts outside the limitations of 'group think'.

The critical point to note here is that when a group (national, social, religious etc.) higher purpose is perceived to be no longer in line with personal values, when individuals or sub-groups believe they have given up too much to belong (subordination no longer delivers for them) then the group will be challenged.

Defining Difference and Risk

Our reality inescapably determines our values, our purpose and telos. It provides the framework through which we perceive everything; it defines us and gives meaning to our lives. Driving our ambitions, hopes and dreams, our reality, the objective and artificial mix of our experienced world, provides the context of our lives.

Understanding that each of our 7.4 billion fellow humans experiences a unique reality is not a trivial issue. Clearly, similarities exist, groupings reflect commonality; nevertheless, no two people experience life equally. Personal ideology, beliefs, what matters to us and drives us, all reflect our reality. In future chapters it will become clear that emerging technologies are about to severely challenge and alter our realities. These technological developments will lead many to question and re-examine just what it means to be human.

Regardless of our particular reality, we all face common existential risks. Whilst these vary in impact and consequence we share the threats of natural disasters (solar activity, meteor strike, earthquakes etc.). Our realities do however result in a range of diverse risk profiles. It is clear that a nomadic tribe will not be hugely affected by the loss of electricity, telecommunication or a stock market crash. Those same events however would cause massive disruptions to an advanced society.

Consider the following risks from two perspectives. How would they affect someone living in a 'grounded' objective reality, say a tribal African hunter? How do the risks differ to someone living in a high density modern city environment?

Common existential planetary risks

Food, water, shelter (shortage, loss of supply)

Energy supply failure

Loss of health services, medicine

Resource deficiency (economic crisis, distribution and access disruption)

Breakdown of law and order, loss of security

Technological failure (prolonged loss of ICT)

It's self-evident that modern societies have become fully technology dependent. Anyone living in a high-rise apartment should at least consider the loss of electricity. How would people cope with a loss of power (both supply and back-up) if they are living on the 25th floor? When electrical systems like elevators and even lights cease to function will they cope? I believe most couldn't even make the trip up the stairs. Equally what if something basic like the supply of water was stopped? Could the average city dweller even find water and actually get it back home? Where, in a city of several million people, would they go?

I'm not trying to panic anyone but should we not at least consider these risks and think about mitigation? The point that needs to be understood is that our risks do depend on what our precise reality is. The fact that our current critical technology dependencies will in the medium term only grow stronger should be a concern.

Grouping and Fracturing

In Chapter 2 we considered the formation and groups for various reasons of advantage. Today's groups, nations, have largely formed because of geopolitical and geographical dispositions which at least at the time of their formation had a compelling purpose and could be thought of as 'homogeneous'. As these groups mature and absorb change over time, they form specialised sub-sets and diversify in their purpose. When such development is gradual and organic then they are accommodated and embraced by the larger group.

With exponential technology 'rapid change' is enabled and the group's ability to absorb this and the resulting inequality of access becomes strained. A once homogeneous group now exposed to ICT enabled disruption will accelerate the group into a severely fractured state. Whilst this may seem counter intuitive (ICT connects and shares) it is evident when we consider individual's education, intellect and access determined by wealth, means and opportunity. There will be a greater diversification and the growth of difference. This is evident in most 1st world countries today.

Superficially, technology has apparently connected all and made everything easier. Is that really true? Most modern societies have massive populations of people 'left behind'. These include the elderly, the economically disadvantaged, the poorly educated, the redundant workers and their families. Whilst there appears to be a perception of smart societies the level of technological understanding has never been lower.

The internet, the ability to 'Google it' continues to be an argument used by too many – that they needn't learn or know this or that because if they need to know – "I'll just Google it". Socrates[20] bemoaned the development of writing. He feared that "they would receive a quantity of information without proper instruction" and that they would "be thought very knowledgeable when they are for the most part quite ignorant". What would he say about today's internet dependency? This too is accelerating the fracturing of groups; it establishes differences caused by diverse reliance and understanding of technology.

The above becomes critical when we consider governance. It becomes almost impossible to govern large groups when individuals live in vastly differing objective and artificial realities. This is evident around the world today. Additional differentiating impacts caused by technology, economic, geopolitical and environmental factors continue to create massive diversification and are the cause of destructive inequalities.

Humanity's Grand Challenges

I mentioned Singularity University in my opening chapter. Whilst SU is not unique in its aims to solve global challenges through the development and application of technologies, it is rapidly becoming a major influencer and it's useful to detail what it has identified as our greatest challenges. SU lists the following as 'humanity's grand challenges':

Food – sustainably supplying the nutritive needs of the growing global population

Energy – reliable, clean energy sources

Water – ensuring safe, reliable drinking water

Security – protecting people and infrastructure from immediate dangers

Health – basic healthcare access and preventative care and support of medical advancements

Education – personalised, lifelong learning

Environment – sustainable environments and earth systems

Poverty – ensuring socioeconomic and basic needs are met

Space – addressing global needs and threats via space exploration

Governance – critical, recognising that technology alone is not enough

Disasters – mitigation and response

Given that there appears to be general agreement on the above challenges it's a bit of a revelation that our technocrats are seeking to develop technologies to address them. Isn't it somewhat obvious that we've had the technologies to solve most of these problems for decades? It should be apparent to even the most casual observer that it is not the technology that we lack but the will and intent to actually apply it to these issues. This is the reason that 'governance' was a recent addition to the above list.

Governance does indeed address what are perceived as challenges; usually these are the challenges that the governed population focuses on and cares about. A recent survey in the US[21] revealed that the most concerning issues for Americans were:

International terrorism

Development of nuclear weapons by Iran

Cyberterrorism

Global spread of disease

Syria conflict

North Kora's military power

Refugee influx into Europe and North America

Global warming and climate change

Note that this list does not include food, water and poverty – some of the most disturbing and easily remedied challenges. The concerns of citizens reflect the media guided group norms and government focus; they do not in general reflect the real global challenges. It should therefore not be surprising that attention flows to the areas of group concern rather than to what global priorities ought to be.

We live in a diverse range of realities. Our differences are significant. We acknowledge humanities grand challenges but do not address these as matters of common priority. Technology will undoubtedly provide us with better tools; tools we've so far not applied effectively.

The fracturing of groupings is divisive but a logical outcome of our realities. We are to varying degrees becoming highly dependent on technologies to sustain our high density groups. This dependency is self-fulfilling; the more we develop and adopt advanced technologies, the more dependent we become and the greater the need for ever-better technology.

CHAPTER 5

The Global Context

Technology alone is not the Answer

In Part 2 we will consider various technologies and the advances in scientific research and development they continue to enable. I will explain that these powerful exponential technologies are, among obvious developments, also leading to unprecedented hyper-connectivity, information sharing, cooperation and the emergence of the 'crowd'. An ever growing crowd of connected, online people that are at once, consumers, advisors, informants, investigators, innovators, judge and juries of everything that is being said, done or produced by individuals, corporations or governments.

The potential 'good' that these technological developments can help create, and of course the associated risks need to be examined in context. Many are valiantly spinning the positives; few are considering the inherent risks.

There seems to be general agreement on one very significant fact: technology alone, exponential or linear, cannot bring about any abundance for all nor address serious emerging global problems. Issues that are problems to some and opportunities to others; issues that are increasingly becoming evident include the geopolitical transfer of power and influence to the 'East', collapsing economies, rising inequality principally in the developed world, concentration of wealth into the hands of fewer and fewer people, wealth drawn from shrinking middle classes are just some problems to be addressed. I will detail these below because 'leading into the future' requires us to understand the situation clearly.

In reading the pages that follow ask yourself what it will take to fix any issues that you recognise as being a problem, a global challenge. Consider that we do live at a time that has already commenced to be severely disrupted by new technologies. When I go on to detail exponential technologies you may wish to note their potential to address global concerns and you will need to decide your position; make a judgement call on whether you believe that the world's population will indeed benefit from them or destroy itself before potential solutions can be implemented.

I mentioned earlier that I attended one of the executive programme courses at Singularity University in Silicon Valley (Dec 2015). On the first day, we, the eighty or so participants from twenty-three different countries, were invited to arrange ourselves along a line marked out on the floor with the most optimistic at one end the most pessimistic at the other. We were asked to base our optimism or pessimism on our general feelings about the global future being positively affected by exponential technologies. Unfortunately I was a bit tardy entering the hall and I headed towards the pessimistic end. Before I had the chance to walk up the line towards a perhaps a more optimistic position I was handed the microphone as the most pessimistic person in the room and asked to explain to the optimists why they were wrong. I said something like:

"If you think that technology is going to solve the world's problems then you're dreaming. We've had the technology to feed everybody on the planet, to provide water and sanitation, housing etc. for decades. It's not technology that we lack it's the will to apply it where it is needed. The answer is in governance: moral, ethical governance that distributes [resources, assistance, health care and technology] equitably; governance that actually cares for humanity, all humanity".

Having on the first day designated myself as the sceptic was not my intention. Surprisingly, many of my course colleagues approached me during the days that followed and wholeheartedly agreed with what I had said and the position I took. So why were we there?

I learnt that most of us see the potential and the inevitability of exponential technologies. Now I'm absolutely sure that the technology is unstoppable. Life for all of us will change drastically; also at an exponential rate. What many of us now know is that these changes will bring extraordinary beneficial possibilities and with it equally extraordinary risks, risks that can end the human experiment. If this seems too fatalistic to you then read on and decide your position.

Usually when the up or downside benefits of technological changes are discussed it is done in the context of a utopian 'make believe' world. A world that is interpreted not as one of difference (the 7.4 billion individuals each with their own perceived realities discussed, in chapter 3) but one taken as if it were a homogenised world of equal needs and desires. Discussions rarely put this into a real life, real time context; healing the sick, providing sanitation, food and water whilst at the same time bombing nations, imposing crippling sanctions, providing financial aid with 'no win' conditions, exacting non beneficial trade terms, exploiting poorer nations' cheap labour conditions, perpetrating terrorism or supporting despotic regimes is not encouraging and certainly not an equalitarian environment ready to benefit all through technology solutions.

Developing advanced viral responses in laboratories is well and good but when the world lacks an appropriate and timely response too many are simply not treated in time; Bill Gates speaking on the recent 2014 Ebola outbreak[22] said "There's no need to panic... but we need to get going" and invited us to plan, target research and train health workers. There was no mention of any lack of technology – the clear lack was in a coordinated global response and a lack of appropriate decision mechanisms, commitments and action from the very organisation that we mistakenly believe to be in charge, appropriately resourced and ready – the World Health Organisation.

The extent to which any human technologically driven benefits are realised and distributed will depend on our management of not just the technologies but of every aspect of the global community. That means we need to manage the world's nations, the geo-political relationships, all in a sustainable social, environmental and economic context. Right now, that seems like 'Mission Impossible'.

Technology and the interconnected world today has already enabled and empowered huge numbers of nations and individuals to bring about global catastrophic events. Never before in our human history have so many been able to spread their influence, to act on their desires, good or evil, so easily. The technology to cause mass disruptions is already here and growing as you read....

I'm tempted to provide a list of the many options that a terrorist or criminal organisation, a rogue nation or some lone lunatics have at their disposal. There are thousands of tech-enabled attacks readily implementable – just speak with security operatives or read Marc Goodman's[23] excellent but disturbing book – 'Future Crimes'.

The means and methods that could cause global carnage and result in outcomes that could severely disrupt or destroy humanity are readily knowable and doable. If I did detail these options, it would be one more 'how to' that we don't need. Of-course I would also have to add to that long list all those actions that nations, the official nations that sit at the UN table, could take, either alone or in partnership. That list would include manufactured regime change backfires, black operations and false flag operations gone wrong, trade related conflict, all out limited or full nuclear war, again - etc.

I've read the optimistic if somewhat illogical statements that seek to reassure us that "the world has never been safer than it is today". Calming but unfortunately not convincing. I've often been asked whether I believe in the fundamental 'good nature' of human beings and I've usually said 'I'd like to' but I'm having real doubts. When I see what we've done to each other, when I consider pervasive greed, astounding ignorance, prejudices, hatred, inequality and passive tolerance of totally unacceptable human rights violations, I find it hard to be very up-beat about human nature.

When I first became aware of the 'abundance for all through technology' concept espoused by the movers and shakers primarily based in Silicon Valley, I was sceptical to say the least. I had no issue with exponential technology – that's simply evidenced and somewhat predictable. What concerned me then, as it still does today, is the knowledge that the world is simply not governed well enough to ensure that the promised abundance will be available to many; certainly not all. My point continues to be that we have an abundance of technology now that could assist billions of people - so why is this not happening?

At the 2015 World Economic Forum in Davos, Switzerland, global leaders including Pope Francis (the Pope), Christine Lagarde (MD IMF), Jim Kim (President World Bank) and Mark Carney (Governor Bank of England) were "among those who stressed the need to ensure the fruits of economic progress are shared more equally" (the Guardian, Australia). That's reassuring and it is even more reassuring that Mark Carney took part in a panel that was titled 'a richer world but for whom'.

A realistic appreciation of the global 'state of play' is needed to accurately begin to define what will need to change to ensure that the benefits of exponential technological advances will indeed address the global challenges we all face. We need to suspend personal, social, national pride, prejudices and protective attitudes and with a no-blame mindset, honestly define what needs attention. We all understand that avoiding real and even perceived issues will not correctly define the problems to be solved.

There are two key challenges that must be addressed. Firstly we need to establish an effective and peaceful coexistence for all. A global governance system that will ensure sufficient stability and one that treats sovereign nations fairly and equally. It is preferable that this be understood for what it represents – an urgent preventative step to avoid nation based revolutions and uncontainable global conflict.

The second challenge is to escape the 'might' and 'trade' related practices that are the foundation of domination, inequality and exploitation. Technology can make abundance for all a reality but we will need to fundamentally change economic and social paradigms to create a world ready to share. This task is extremely difficult and is most likely not achievable until the people of the world have suffered so much that they are prepared to accept change based on the mutual need to survive. This is just human nature – nobody wants to give anything up or be helped until it's too late.

I need to be very clear that the following pages are not sugar coated. The world has serious issues to resolve and to the extent that it frames the world to be lead in the future and a world that is additionally evolving form a biological evolutionary history into a self-engineered evolution it must be considered in a truthful if uncomfortable light.

See it Truthfully

Let's suspend any self-imposed fantasies for a moment. We in the 'West', the 'Anglo-celtic' derivative populations, the mostly self-deluded, are proudly observing our favourite religions. I'm not talking about any brand of Christianity, no I'm talking about the twin fantasies of democracy and the free capital market. How ignorant and isolated from fact would we need to be if we continued to believe that these two stupidly sacrosanct pillars of our artificial world are real in any practical sense?

Democracy, and I'm talking about the Greek definition of 'demos' – the common people as opposed to the Magna Carta derived versions that focused on the democracy for masters, (the kings of the old empire). In the pages that follow I will describe some issues that should establish that democracy in perhaps the staunchest proponent of it, the United States of America, has failed; if it ever actually existed. The argument is presented to you not to bash America; the people don't deserve that, but to invite you to consider democracy in the light of how it actually is in the lucky, decent and admirable land of the free? This is the regime to be forced onto everyone else - really? Why would anyone want that?

Another rant: *The most shocking realisation however comes with the knowledge that the elite, the powerful and those in charge (not necessarily those in government) know this all too well. It is the general population, the 'sheeple' of the western World that is largely ignorant of historical facts, misunderstands reality, is fundamentally unintelligent and poorly educated, mostly leads a trivially focused existence.... that believe the myths. The myth of democracy, that 'our way' is the 'only way' and being so proud of their own compost, is held with a conviction that only those that are "kept in the dark and fed on bullshit" (political and media cooperation in action), the mushroom sheeple that dominate our masses, can hold.*

The economic reality is no different. Free capital markets do not, have never and unfortunately probably never will, exist. Of course they exist in text books and in theory lessons; just like democracy, but in reality, in the practical and evidenced reality they are but smoke and mirrors. Again I will hold up some evidence to justify my assertion. How can a globally dominant free capital trade system, in fact simply 'capitalism' that is heralded as being so uniquely good for all, result in the inequality, widespread poverty, subjugation and ruination of billions of people? How is that possible? If capitalism is so great, why is America in so much debt? Why has the EU crumbled taking every member nation down with it? If that is not obvious to them it is because it is unpleasant to confront your own demise and collective incompetence. The 28,000 highly paid bureaucrats that live the EU management fiasco in Brussels can't admit to something like 'we've failed our people'. How could they? Was there any time in our history where a dying civilisation recognised its own downfall and avoided it – no!

Again, consider the reality, a reality that unless you are totally immersed in a virtual reality or trivial life you must be putting experiential evidence, the daily news clues and headlines together – right? I can comment on the majority of our human colleagues that are the sheeple – because they wouldn't ever read this book (or any other for that matter) and they don't watch any factual news (in our western world we don't get actual 'news'; we get sound and video bytes that are selectively edited, sensational and short lived).

There is an obvious need to understand the real global situation. By any measure, the world has serious, very serious, issues to contend with. Note that today's state of uncertainty will not magically resolve itself. Answer or consider these if you can:

- The refugee crisis affects only Europe (and in particular southern Europe) and this will be resolved when...? How?
- The Middle East will become a peaceful region and will not have a destabilising influence on the rest of the world. Terrorism will not be fuelled by bombing nations into the Stone Age and the radicalising of ideologically different ethnic groups within western countries will cease by itself.
- The weapons of mass destruction available to terrorists, friendly black operations, rouge nations and criminals will naturally become less and less lethal and not used frequently. [If you believe that then I have a bridge that looks like a big coat hanger (Sydney Harbour Bridge) that I'd like to sell you!]
- The poor, the impoverished people, the currently silent if moaning sheeple of the old 1st world will continue to wither and accept that they are destined to be irrelevant, redundant and enslaved by the few. They will never rebel, overthrow the powers that be; they will just go away and we needn't concern ourselves with them as long as we give them trivial apps, social media and let them be 'pretend' home owners – or at least aim for that. All good; now for that trade deal...
- Global warming – so what; it would need to rise by hundreds of meters before it reached my holiday home in the Alps. 100 million refugees from low lying nations – we can handle it, just send them off to Europe – they're experienced in that. The islands - who cares, after what we did by repeatedly nuking the Marshal Islands (poor buggers still suffering) a bit of water – so what?
- Global peace - I hope not! Our massive defence industry depends on conflict and we'll kill anyone that doesn't see it that way. Stuff the world – we have business to conduct – and they are!
- Learn from the past. We benefited enormously from WWII; we created the German powerhouse and afterwards the EU, we

created the Japanese manufacturing empire; we won the cold war; we need another World War; it's the best way out of our current situation…relax we're working on something similar; we'll destabilise any competing pretenders to global power, fracture nations…you'll see; they will beg us for help.

I'm sure you get the picture. Of course I'm trying to hammer the point. Where I'm stuck is that isn't all of this so obvious? Do the majority not see the futility of war? I'm not naïve, I've served in the military; I understand that sometimes force is the necessary and only option. I've seen this in reality. When a terrorist isn't interested in negotiation, like a suicide bomber; there is no choice – act and act quickly – but war for regime change? War for commercial advantage or to keep corrupt or incompetent systems going; to enrich individuals - I don't get it.

Clearly, the world is already changing and developing at a fast pace and with the additional, imminent impacts of exponential technologies we are in for some exiting times. Just how exciting it gets will depend on us and specifically how we act in the next decade.

You might, after considering the likely consequences of new technologies determine that we are not ready for them, that indeed our liner minds cannot contain or control their impacts. You might feel that their potential to benefit mankind is balanced or even outweighed by their potential to be used for evil purposes. The one issue we should however agree on is that we had better understand the world as it is now; as it really is as opposed what we might want it to be or what we have been led to believe without conscious consideration.

If you were a corporate, political or social leader how would you brief yourself? What would you insist on knowing before you exercised your leadership, your influence?

I believe you would start with past performance. What worked well? What didn't go so well? What were the objectives and were they met and to what extent? What is the residual perception, reputation and who 'knows, likes and trusts' your organisation?

Next you would examine the current situation. You would ensure that your strategy and enabling objectives were correctly aligned and targeted, were appropriate to meet objectives and that their implementation was effective. You would want to get the true picture; all strengths and all weaknesses. If you were a serious leader you would insist on the whole, factual and objective truth and you would seek it with a no blame mind set. Whilst you might need to excise unacceptable past practices you would of course focus on how that organisation is positioned to move forward.

Only when you are fully aware of the state of play, warts and all, could you clarify the task at hand and set about leading it into the future.

I believe that the world is poorly governed; and that's an understatement. The noble aims of exponential technology providing abundance for all is absolutely not achievable if we don't sort out the global nightmare that exists and is rapidly getting worse. I'm not advocating a one world government or a utopian 'Pollyanna' world; rather, I urge the establishment of sufficient global governance that yes, provides for a peaceful coexistence for all. Only then can the aims of 'abundance for all' begin to be realised.

No doubt you have your thoughts, your own assessment of the world today. None of us can claim a detachment from the nightly news that informs us on the latest poverty figures, famines, disease outbreaks, crime, terrorist actions, human rights violations, the global refugee crisis and the updates of this and that war. Although these news programs are liberally sprinkled with upbeat stirringly patriotic words from our political leaders there is a growing awareness that they are collectively simply not addressing the real challenges.

In the pages that follow I would like you to read them as if you were a global leader or key advisor. What would be your conclusions about the past and recommended actions for the future; just as if you were the 'boss', the manager, CEO or president, briefing yourself as mentioned above.

I'm necessarily going to be a bit general and where you disbelieve or strongly object to the facts I'm presenting, please do verify those issues for yourself. We do need to be factual and objective. Those of us who live in the 'Western' developed world need to have a good hard look at ourselves, what we have done, well or not so well. We need to realize that we have no claim to any form of superiority over anyone; that 'might is right' is no longer acceptable and that those 'underdeveloped' nations are finding their voice just as past empires are re-emerging.

The Small Print of History

At the September 2015, 'Fourth World Conference of Speakers of Parliaments' a consensus was reached on the importance of learning from the past, to reflect on history and to looking to the future in order to create a better world. Zhang Dejiang, China's senior legislator (Chair, National People's Congress) put it simply: "What happened in the past, if not forgotten, can serve as a guide for the future." To understand and provide some insights into what is dominating the nightly news headlines we do need to reflect on the past – it has after all created the present.

I recall when I first learnt about a horrific terrorist shooting in Tunisia. The latest count was 39 dead and 36 injured following an attack on mostly European tourists enjoying the here-before peaceful seaside resort at Sousse Beach. Tunisia, that small North African nation hugging the southern Mediterranean Sea between Algeria and Libya – why?

At the same time I heard about another terrorist attack in Kuwait City. A bomb had been detonated in the Shiite affiliated Ai-Sadiq mosque killing at least 27 worshippers and wounding many more – why?

The above breaking news was soon followed by the now all too familiar 'claim of responsibility' by the Islamic State of Iraq and the Levant (ISIL); [note that the Levant is the historic term for the Middle East region now defined by Syria, Lebanon, Israel and Jordan (general area of crusader objectives?). ISIL is also termed Islamic State of Iraq and Syria – ISIS or Islamic State – IS)]. Again I asked myself – why?

But I knew why and it's not a comfortable awareness. Fuelled by the outrage that followed the 9/11 (2001) attack the US Defence Secretary Donald Rumsfeld was engaged in planning the decapitation of the Iraqi government.

Rumsfeld's "secret memos"[24] (based on a declassified talking points agenda) revealed some disturbing considerations. It's not difficult to conclude what their thinking was when it has been revealed that they thought about and discussed various justifications like: 'how do we start a war', what if we say 'US discovers Saddam connection to Sept. 11 attack or to anthrax attacks?" and 'disputes over Weapons of Mass Destruction (WMD) inspections?' and 'start now thinking about inspection demands'.

Planning proceeded, arguments were presented, Iraq was demonised, the US and its allies, principally the UK, Poland and Australia had made a case to convince the world that Iraq possessed WMD and we watched as Iraq was invaded and occupied in 2003. Eight years later, the US withdrew from Iraq. There has been far ranging and continued criticism about all aspects of that regime change from the illegality of the invasion, the inordinate number of civilian casualties (over 105 thousand), inadequate post invasion plans, the folly of disbanding the remaining military, the resulting environment that spurred ethnic cleansing and the total misjudgements and arbitrary nature of the US war on terror. The fact that Iraq was basically bombed back into the Stone Age didn't get much coverage but the destruction of basic infrastructure continues to cripple Iraq today.

The outcome of that, unfortunately not isolated act of judge and jury justified regime change, may not be the single cause of the regional instability that allowed ISIS to create a Caliphate (the precondition for the establishment of ISIS) and pursue its religious imperative to operate towards the anticipated Armageddon (their extreme interpretation of the Quran which is disputed by most followers of the religion of Islam). That particular action, the 'liberation of Iraq' hasn't actually helped stem the spread of terrorism – has it? One observation however is pretty safe. Had Iraq not been invaded, ISIS could not have formed; there would have been no vacuum, no ISIS territory for a caliphate and they certainly could not have captured the massive amounts of military hardware and munitions that they took from the destroyed Iraqi Army. ISIS was enabled by the destruction of Iraq.

Clearly ISIS is engaged in a horrific implementation of its objectives. Their fanatical beliefs require them to fight to the last 5000; it compels Muslims with a similar interpretation of the Quran to travel to the caliphate, their lands, and to join their struggle. Their religious belief does not allow them to recognise boarders, to negotiate with anyone and makes strict demands on taking lives and territories – it is a self-defeating objective that will, according to their interpretation, culminate in a final showdown between the last 5000 surviving combatants and the enemy (now considered to be the US, alternative interpretations suggest this to be Turkey) and herald the end of the world. Label ISIS as you will, they represent the ultimate enemy; one that cannot be negotiate with, one that wants to bring on Armageddon and relishes dying in the process.

The world needs to acknowledge ISISs unique motivation – they are not conventional terrorists they are religious extremists and nothing short of the removal of their reason for being, 'the caliphate', will stop them.

So Iraq is certainly one of the 'whys'. There are many others. Palestine, Afghanistan (Taliban), Libya, Yemen, Syria, Saudi Arabia etc. are all nations that are generating terrorists based upon perceived and very real injustices perpetrated against them by Western powers. I don't think I need to spell out more terrorist making activities. Political interference, war, bombings, drone strikes, bullying, manipulations and imposed regime change are all creating new resentments every day.

Can we not understand that if someone's innocent child, parent or sibling is killed through an aggressive act that they might be resentful, that they might want revenge. Those statements about collateral damage or incidental, unavoidable civilian casualties don't remove the pain and anger. Should they be thankful for the destruction of their country; for denying them to develop at their pace and reflecting their values? Do we blame them for not wanting to enjoy the nightmare of inequality and rising poverty that the US and European based nations are about to endure?

On an almost daily basis I hear the most incredulously stupid question of all from the lips of citizens of the developed world: "Why do they hate us?"

The modern era of terrorism is awful and we should come to grips with the fact that bullets and bombs are not the answer. We can agree or disagree with its recent origins but guess what; there are some historical and past cultural misdemeanours that contribute significantly to providing clues as to why we might collectively be hated by so many (of course I'm working on the assumption that terrorists don't generally attack, maim and kill people they like).

Our educational institutions don't often teach the awful truths. Not only does the 'victor write the history' the victor also interprets it for their future generations. So what are some of these inconvenient histories?

Let's start with the holy crusades - those Catholic Church sanctioned military campaigns that sent crusaders to reclaim the holy lands in and surrounding Jerusalem. Over 2 hundred years (from 1096 to 1285) European countries dispatched several hundred thousand crusaders to pillage and conquer in the name of God. To many, being a soldier for Christ was the ultimate proof of their devotion to God. Is it likely that the Muslims of the 11th to 13th century could have perceived the crusaders in a similar way to how we view ISIS today?

The West, even today, marks these crusades as significant to the expansion of trade in the Mediterranean and a sort of coming of age. The crusades have been credited with establishing Europe as world hegemony and a catalyst that brought on an age of great literary, spiritual and artistic traditions. Unfortunately, as acknowledged by many historians, the crusades also entrenched Islam as the ideological enemy of the West. This ideological identification of Islam has continued and it is well understood by the many nations that suffered at the hands of the crusades. It is a cultural reality that is an integral memory; it is in the DNA of the people of the eastern Mediterranean countries, the citizens of Istanbul (sacked and destroyed by crusaders) and the North African countries that were traversed by the Jerusalem bound soldiers of God.

Their recollections of history and the perceived wrongs of the predominantly Christian invaders are naturally different to our accepted and more convenient history. What we see as exploration and a spreading of our values they might interpret as interference. All of that is in the past. No one alive today can be blamed for the actions of our forefathers so it's all forgotten now, right? Well that's what I used to think too.

By the 15th century, sea faring technology had ushered in the era of colonisation. This enabled the great European expansion, the modern era of colonialism from about 1500 to 1900. Western European nations (Spain, the Netherlands, France, Portugal and England) took to their vessels and criss-crossed and explored oceans in the search for new worlds; worlds that might provide new economic benefits or immediate gains.

Where these European conquers encountered native civilisations that were 'undeveloped' they labelled that territory as an empty land or 'terra nullius'; as was the case with Australia. Where they encountered a level of complexity in the newly encountered land they would seek trade or enriching opportunities. In either case, new territories were claimed, fought over and often settled.

There appears to have been little consideration for the native populations encountered. Colonisation, by design or default, became the mechanism of acquiring wealth, territory, commercial expansion and hence power that resulted in the subjugation, absorption and assimilation of encountered cultures into that of the colonising country.

Colonisation spread slavery through the practice of settling slaves from the country of origin or the enslaving of native indigenous people. Don't need to spell this out – we all know this.

Of-course I need to highlight that at that time, just as now, politicians, rulers, the sponsors of expeditionary forays were all spreading the benefits of their own most valued cultures. Just as now they would utilise their superior technologies and trade networks to dominate and subordinate the underdeveloped; spreading peace, and good will among all they conquered.

All forgotten now, and in any case we did bring cricket to much of the world, those South Americans didn't need all that gold and silver, the Chinese couldn't drink all that tea and the natives of the empty lands – well, what were they going to do about it?

Those European countries were still in the service of God. They knew that their god was the only one; they were superior by any measure, they had the right to spread their beliefs, their culture and governance models to all. Unfortunately not all encountered belief systems were eradicated. Some survived and developed. Some developed quite unreasonably; some of these new territories, despite subjugation, resisted European domination. Those ungrateful Americans actually kicked out their masters in the war of independence.

The rape and pillage of people and foreign lands (sorry that's how the modern citizens of past vassal nations might view colonisation) still reverberates around the globe today. The apparent smile and pleasure at diplomatic interaction with their past masters is often just the demeanour of a past servant; free but still a little wary of new gifts, advice and opportunities being offered.

I have often heard of the noble intentions of the colonising powers. They admit to exploiting trade for their own enrichment; fair enough – but it was one sided and self-serving; quite understandable but let's own up to that and not pretend they didn't know.

Consider Lord Macaulay's address to the British Parliament on the 2nd of February 1835:

> "I have travelled across the length and breadth of India and have not seen one person who is a beggar, who is a thief such wealth I have seen in this country, such high moral values, people of such calibre, that I do not think we would ever conquer this country, unless we break the very backbone of this nation, which is her spiritual and cultural heritage and therefore, I propose that we replace her old and ancient education system, her culture, for if the Indians think that all that is foreign and English is good and greater than their own, they will lose their self-esteem, their native culture and they will become what we want them, a truly dominated nation"

Way to go Lord Macaulay!

I was given this in the form of a jpeg image of an old newspaper clipping by a past engineering colleague; a young, intelligent, Indian migrant to Australia who kept this on his iPhone. What does that tell you?

I acknowledge that colonisation, transportation of undesirables, relocations and foreign occupations were indeed global and not limited to Western European actions. I'm just picking well known cases to make the point that we cannot ignore the facts and whilst it might serve our purpose to tell it from our point of view it is not necessarily how the previously silent world recalls and acknowledges those episodes in history.

So one more example: the opium wars. England had established a massive opium production industry in India and coveted Chinese tea and porcelain (they do love their cup of tea). The Chinese, the Qing dynasty, refused to trade these commodities for English trinkets and goods, they wanted payment in silver. The English had the solution; sell opium to China for silver, buy tea and porcelain with that silver – simple. That worked well and by the late 1830s, China had millions of opium addicts and suffering the health and social impacts of this mass addiction.

By 1839 the Chinese had had enough and seized and destroyed an estimated 1,400 tons of opium. This sparked the first opium war (1839 to 1842); England's superior naval power inflicted a humiliating defeat on the Chinese. China was forced to pay hefty fines, open up several ports, lease Honk Kong to the British for 99 years and afford British citizens special rights in China. The situation that that humiliation had created was made worse by Britain's constant seeking of further gains. A minor event, the boarding of a British ship by Chinese officials arresting several Chinese crew members in 1856 was enough for the British, now joined by the French who also had aspirations for China, to engage in the second opium war. Again, European technology triumphed and by 1860 the Chinese were forced into a new treaty. New territory, the Kowloon Peninsula, was added to the Hong Kong lease deal and the trade in opium was legalised.

All forgotten now; again, not really, although one might expect a certain shame for forcing opium trade on anyone. That ugly history told by the victors as simply enforcement of free trade policies, a protection of its international merchants (today's corporations) and a proud dominance of weaker nations (India and China). To the Chinese, these wars weakened and humiliated the Qing dynasty and left China open to significant foreign influence. Their history chronicles the opium wars as the cause of a century of humiliation at the hands of European powers; a century that lasted until the 1949 Communist revolution, the War of Liberation that heralded the Mao Era.

Having had a brief glimpse of some of the above events, can we not acknowledge that they have shaped our image and defined our nature in the eyes of today's emerging nations? Does any of the above resemble the way we engage with nations today? Are we still manufacturing and exploiting conflicts. Are we still exporting our values at the expense of foreign, indigenous, native values? Do we still act in a manner that assumes and broadcasts our superiority and conviction of entitlement?

I introduced this section by referencing the June 2015 terrorist attack in Tunisia. Clearly this was a shocking attack and many innocent people were killed and injured. The resulting global condemnation of that and similar attacks are of course totally justified and we must all do everything we can to stop that sort of cruel and senseless mayhem.

Given that there were 30 Britons among the dead it was entirely appropriate for the British Prime Minister David Cameron to express his outrage and to call for fight back. The Telegraph (article dated 29 June 2015) reported that "… the Prime Minister says that we can only defeat terrorism by promoting the British values of "peace, democracy, tolerance and freedom"…and that "We must be stronger at standing up for our values – of peace, democracy, tolerance, freedom," he says.

I totally agree with his sentiments and invite everyone to join the fight by acting precisely on those values. Further, our actions in the past, regardless of motives, need to be accepted on a factual basis. We build on what we actually did, what we created and not what we would like to think we had done. What is comfortable is irrelevant. It's as simple as 'man up and move on'.

In Australia we've had to man up. It took us an embarrassingly long time to say 'sorry' to our indigenous people. Our Aborigines had culture, traditions and values when Captain Cook claimed this 'empty land' (Aborigines fought valiantly but in vain) for Great Britain.

Our 'sorry' hasn't changed the many injustices and cruelties we modern settlers had inflicted on our native people, it didn't rewrite our history. Many continue to sincerely implement a range of measures to rectify our early mistakes and I have to admit were not making sufficient progress. We can't undo much; most of us can't hand the land back and we certainly can't go back to our countries of origin. We have to find solutions.

What we did achieve however, in our national 'sorry' is that we acknowledged a persistent hurt. We asked that the actions of our forefathers be forgiven that the natural kindness and grace that our proud Aborigines have in abundance, allow us to embrace, to unite and to pursue a better future; one that recognises the wonderful diversity and fragility of life.

Global Governance - Where is it?

Given our current and exponential growth of inter-connectedness and growing inter-dependence, there is little point in looking at the micro level, the individual, without first understanding the big picture; there's no escaping the fact that we share the one planet. I'm also working on the assumption that visionary space enthusiasts, whilst desiring the exploration of space for resources and alternate human habitats aren't actually planning a viable escape route for all of us. I mean, picture it; long lines of quietly optimistic modern refugees (about 7.4 billion people), loaded up with 'must have baggage' waiting their turn to board 'Virgin' flight 101 to Mars – the new dream of salvation and abundance for all. Yeh, right!

For now, before we think about racing off to another planet we should probably try to save the one we're on and until we come to at least a general agreement of what our global 'state of the planet' is, we clearly can't address any necessary changes.

Much has been written and said about the emerging world order. There is no shortage of intellectuals busily labelling and categorising our geo-political world and much is changing. From the comfort of the bi-polar cold war, a clearly divided world where we all knew to cower in our shelters safe in the knowledge of a defined enemy, to a mono-polar world following the collapse of the Soviet Union to the present realisation that …wait a minute; what about the emerging and re-emerging new kingdoms? Oh, well perhaps a multi-polar world might have to be next – all has been discussed, and opined upon.

We are living in an uncertain era. Not only is the new world in transition and global powers are re-evaluating alliances, its nations are torn between what has served them well in the past and interpreting perceived emerging opportunities and threats. This change is made so much more significant and daunting because our established norms, both economic and social, are being altered by the now unstoppable age of 'hyper-connectedness'. Whilst political and industry leaders are valiantly trying to govern their charges, the social global family is becoming restless, inquisitive, informed and basically very difficult to manage. Like a rebellious teenager, people everywhere are questioning, resisting the status quo and demanding justifications that make sense today. Empowered by exponential technologies, the desire for immediate solutions, no lingering memories, 20/20 hindsight and little respect for past sacrifices or stupidity, they demand it all: perfect governance, total freedoms, equality for all, a pristine planet and (mostly) for others to take the risks and create the utopian world they believe they deserve.

A utopian world order is a worthy aspiration. Perhaps our leaders will work to achieve that goal and perhaps being armed with the technology that can indeed provide abundance for all, we might actually achieve it – however we do have some challenges to overcome. The greatest challenge is to rejuvenate global institutions; institutions that may have served dominant nations well in the past (let's be generous here) but were created in and for an era now long gone.

Global institutions now need to re-examine their purpose, restate their missions and reconsider their memberships, operations and member rights to be inclusive and transparent. In short, they need to serve the interests of the multi-polar emerging world. Denial will not do; nor will resistance to change. Those global institutions that have genuinely, or disingenuously, fought for globalisation, now need to define it and step up to the plate; not doing so will not only render them redundant to the new world order but will potentially result in the creation of a new but far more extensive cold war; an economic war between the old and new world.

I believe a quick look at some key governance bodies is relevant in that I want to confirm the global environment that exponential technologies are about to severely disrupt. I aim to make the case that for technology to be applied to solve humanities challenges we must create a peaceful world, not a utopia or a hegemonistically dominated world but one that can set aside serious difference and cooperate for the good of all mankind – and that is indeed our joint global challenge. It is the most difficult and it is potentially unachievable. It is an aim worth trying even if the obstacles loom large.

There can be no logical argument against benign global governance. I draw a clear distinction between governance and government. One world, one government – not likely just yet; what is possible though is an agreed governance system that allows for diversity of governing regimes and respects sovereignty and the right of self-determination. A governance structure that empowers itself to jointly deal with Earth's existential threats and has the authority, power and commitment to act decisively against common threats (and yes, this includes rogue nations, criminal organisations, cyber and traditional terrorists etc.).

We need to collectively make the new world order safe, functional and sufficiently capable to manage and apply to mutual benefit the innovations our people are bringing into the world. Unfortunately, we have to start from here and now; and as I've said before the here and now systems need a bit of work to bring them up to speed. So, how well is our global society managed?

There's no simple answer to that question. You could take the view that it is very well managed to serve the elite; they are growing more powerful and wealthy by the minute; they are the masters of the economy, the global strategists and power brokers. If you however try to judge global management from a holistic human perspective you might justifiably conclude that it is absolutely poorly governed and that it has never been managed well. I say that the world is inappropriately governed.

I say this because whilst we have much government and many supposedly global, primarily western institutions, we do not have any form of truly representational world governing body.

The closest global governance body we have is the United Nations (UN). Established in 1945 (a rebadged League of Nations), the victorious nations got together to make rules for their mutual interests and created a western dominated alliance for self-empowerment. Today the UN still represents an organisation based on the post WW2 environment.

An example of this is the very powerful Security Council which is made up of 5 permanent members (China, France, UK, US and Russia) and ten non-permanent members. The five permanent members have the power of veto based on nothing more than the power groupings of the post war world. Some 70 years later, how is it that France and the UK have such a powerful veto compared to nations like India, Indonesia etc.?

Unfortunately, the UN is incapable of governing the world; that's not its charter or mandate. It is certainly not democratic (demonstrated by selective nation veto powers) because, well that's not it's charter either and sadly it's also not very good at governing the world as a default 'go to' global management come leadership group.

You can study the history, composition, processes and complexities of the UN at leisure – there is no shortage of freely available material and you might then consider how it's performing its default leadership role. I respect much of the good work that the UN has and continues to do. Its agencies are busy around the globe and too many times are called on to deal with situations for which they are not structured or resourced. Being dependent on donor funding, all too often, UN reports, responses, actions and inactions on important global issues are influenced by donor nation self-interest.

Does the UN then function as an effective governance body?

> If you answer '*yes*' - then why haven't the world's problems been addressed? Where's the global strategic plan? How is this plan being implemented and what's our performance against its objectives? Where is the globally equitable democracy, or for that matter, an effective benign dictatorship or a socialist model? In fact, what is the model that the global government reflects and enforces?

> If you answer '*no*' – then not surprisingly we're in for a bad time. Every governed nation in the world will defend the need for self-governance (in whatever form they champion) so how do they expect a non-governed world to perform?

If you're not sure how to answer – well that's actually what we've got; a façade of governance but no evidence of effective global government, thousands of talk fests every year, mountains of threatening to mean something resolutions, agreements, accords and press releases etc. all giving the impression of governance, all driven by various interest groups.

Notwithstanding the 'good' work the UN continues to do around the world, often as first responder, it remains at the mercy of membership funding. It lacks the charter to act; it has no 'teeth' to enforce compliance with resolutions. It does however provide a global forum and valuable research and reporting on global issues. It's a shame that too many of these reports are subjected to nations imposing self-interest based editorial influence.

The UN nevertheless remains a foundation organisation that seeks to coordinate the international community. Its Millennium Development Goals (MDG), the climate change induced shift of focus to sustainable development goals are of-course welcomed and critical but results are mixed. Again, if you care to look up MDGs and their effectiveness you will find a mixed bag of success and failure. There should be no argument that 'good' is being done but here's the issue:

The UN's work (and that of many charitable organisations and NGOs) is like a medical team on a battlefield. Medics doing an outstanding job; saving lives here, there and everywhere, clearly needed and appreciated; whilst the battle rages on around them. The combatants on the battlefield, appreciating and funding these medics are at the same time pouring in more ammunition and working on future conflicts. When the economy of social, political and trade combatants depends on advantage based win-lose, medics like the UN can do little else but render first aid with the only certainty being that the need for such patch-up aid is continually growing.

Other than the UN, what 'institutions' do we have that in some way govern our world?

The Group of 20 (G20, members: Argentina, Australia, Brazil, Canada, China, France, Germany, India, Indonesia, Italy, Japan, South Korea, Mexico, Russia, Saudi Arabia, South Africa, Turkey, UK, US and EU). With no formal charter, this group meets behind closed doors to consider various world economic issues. Not surprisingly, questions are asked by non-member states about their authority to decide global issues; who decided who is a member; why its considerations are not transparent and how and to whom is the G20 accountable. The obvious questions to me are: what's its relation to the UN and why does the UN not act to have the UN General Assembly manage its own, global economic council?

The Group of 8 sorry Russia not allowed to play now 7 (G7, Canada, France, Germany, Italy, Japan, UK and US). If you thought G20 was selective you're in for a treat with the G7. Who decides membership of this powerhouse of finance ministers and central bank governors? The

International Monetary Fund (IMF) decides this based on its determination of the most advanced economies in the world. Note that these are some of the major culprits of the greed driven recent and persisting global financial crisis and the diligent overseers of the banks that have recently been fined billions of dollars for fixing international exchange rates (manipulating currency trading for maximum bank profits). Sorry China, Russia, India, Brazil etc. – clearly you are not required to attend.

The pointlessness of the G7 has recently been demonstrated to all at the recent summit (The June 2015 retreat at Schloss Elmau in Germany) where the G7's Leaders' Declaration listed its commitment to deal with issues such as global economics, financial market regulation, taxation, trade, and foreign policy in terms of motherhood statements conveniently failing to highlight that it was the very members of the G7 and its loyal 'umpa lumpa followers' that were principally to blame for the causes of these ills.

Deservedly, the G7 is described by objective commentators as totally incompetent to solve any global issues and in the words of the UK's Prime Minister, David Cameron (speaking at the UK hosted 2013 G8 summit) "… [The G8] should put its own house in order".

The World Economic Forum (WEF, formed in 1971, based in Geneva)

What started out as a Swiss non-profit foundation has become an international forum that involves thousands of business and political leaders and individuals to discuss a variety of issues along its stated mission as being "…committed to improving the state of the world by engaging business, political, academic, and other leaders of society to shape global, regional, and industry agendas" (Wikipedia)

The WEF convenes several meetings in diverse global locations every year but is perhaps better known for its annual winter meeting in Davos (unkindly labelled as a meeting of ""fat cats in the snow". Like most organisations that purport to solve the world's challenges (WTO, IMF, G7, G20 etc.) the WEF is also accused of promoting capitalistic globalisation strategies that result in the spread of poverty and environmental destruction.

There is fair consensus among observers that 'globalisation' is that particular activity that US, Japanese and EU powerhouses engage in to internationalise economic strategies that they favour and which serves their interest over all else. Sounds bad but would you expect such organisations to manipulate, sorry help, the world to the disadvantage of its members?

Whilst the old Western Powers appear preoccupied with creating global issues and then meeting to solve them, they are failing to recognise that their days of global dominance are numbered and that some highly significant and far more relevant to the imminent future, organisations have formed and are gaining strength. That must be so annoying for them.

BRICS (Brazil, Russia, India, China and South Africa)

This newly formed economic bloc encompasses about forty-two percent (June 2015 statistics) of the world's population. Combining the manufacturing might of China and India with the resources of Russia, Brazil and South Africa will soon dominate the global economy.

There are some key factors that have not escaped the attention of the US/EU bloc or the global banks (IMF and World Bank). The desire for China and Russia to dump global reliance on the US dollar is hardly surprising nor unexpected. The other compelling supportive argument is the spreading realisation that the majority of the world resents the US and European self-awarded leadership role; one that is increasingly being exposed as the self-serving empowerment that it has represented for decades. There is a growing realisation that what has been perceived as bullying through coercion, military threat, political interventions or 'advantage aid' (we'll lend you billions so you can spend it on our goods and services) disguised as international assistance no longer needs to be meekly accepted. BRICS is just one very significant step leading to a global separation from US and European dominance; a dominance that is more like last year's prom queens still traipsing around the world proclaiming their popularity and beauty.

The fact that the old world just doesn't get it; hasn't accepted that things have changed, has forced countries like BRICS members to step outside the old world order. Many nations are "not fully satisfied with the international financial architecture, not fully satisfied with the role that our countries are allowed to have at the IMF and the World Bank" (BRICS Bank Vice President Paulo Batista at the 2015 BRICS Business meeting at St Petersburg).

So what are these nations stepping away from? Simply put, they are cutting the US umbilical cord: that life line that was once the only source of financial and political direction; that imposed on all, its Wall Street model of global financial and commercial management and that Washington centred cord that had arrogated for itself the role of global judge and jury. Perhaps they understand that the US/EU/Japanese domination is over. They are re-establishing their sovereign independence, an independence that is not hegemonistically motivated.

The World Bank, established in 1944, has the admirable goal of reducing poverty but even the US Senate acknowledges some serious issues when it highlighted the following: "The World Bank continues to be criticized about the way that it is governed (always headed by US) by representing the interests of a small number of powerful countries at the expense of the majority of the 188 member nations. It has been described as a pillar of global apartheid". (A Report to the Committee on Foreign Relations, United States Senate, One Hundred Eleventh Congress, Second Session - 10 March 2010)

The **IMF**, also headquartered in Washington was established in 1945 and has a membership of 188 countries. Like the World Bank it also works to reduce poverty. It has a broader charter to also keep statistics, analyse and survey its client's economy and demand the implementation of corrective economic policies that are prescribed by the IMF. Just like the World Bank, the IMF is criticised as a US functionary that imposes on its client's the ideology of the US and dispenses larger loans with significantly more favourable conditions to its allies.

Recent events surrounding the 2015 Greek debt crisis has revealed that IMF's arrogant management is not focused on alleviating poverty.

Rather, in best Shakespearean tradition, this bank behaves like a latter-day 'Merchant of Venice'; a profiting loan peddler. What makes the Greek crisis even more ridiculous is the fact that loans were extended to Greece based on that well proven American winning model of sub-prime lending; a system so transparently self-serving that it beggars belief.

AIIB (Asian Infrastructure Investment Bank). On the 29th of June 2015, 50 founding nations signed the articles of agreement for the banks establishment. Headquartered in Beijing this Chinese initiative is a concrete expression of the new world order. The bank, according to Chinese foreign ministry spokesman, Lu Kang, seeks to "…help developing economies in the region resolve their practical problems in pursuit of common development and prosperity"[25].

With an initial capitalisation of $100 billion, the AIIB is poised to become a powerful global bank alongside the World Bank, the Asian Development Bank, the European Investment Bank and the African Development Bank. All those unhappy customers of the IMF and World Bank now have an alternative and the majority of countries within the stated region the bank seeks to serve have signed on with the major exception being the US (surprised?), Canada (application under consideration) and Japan. What is noteworthy is to look at the non-regional members. Why would these 20 countries want to join? Austria, Brazil, Denmark, Egypt, Finland, France, Germany, Iceland, Italy, Luxembourg, Malta, Netherlands, Norway, Poland, Portugal, South Africa, Spain, Sweden, Switzerland and no 'Et tu Brute?' the United Kingdom.

What has been the reaction from the world's leading democracy, that superpower wanting to rid the world of poverty through its trade policies, political mentoring and its banks (World Bank and IMF)? That exemplar of trade and financial fairness, the US, has questioned whether the AIIB will have adequate environmental and social safeguards and even expressed doubt about the bank's standards of governance. Stop! I think I hear the noise of billions of eyes rolling in unison as this outrageous hypocrisy. Needless to say, the US has been counselling nations not to join the AIIB.

Let's now turn to international trade. There is a proliferation of international trade treaties and agreements. With China's rapid rise to a position of global trade dominance there has been an almost hysterical rush to get 'in on the action'. The US in particular has rushed to encircle China and contain its regional influence by 'signing up' all and sundry to its trade agreements; trade agreements that unfortunately can subordinate nations' rights on any issue that affects profit objectives and expansive ambitions of corporations.

I mentioned earlier that the world is moving into an era of multi-polar global governance. Whilst this is generally acknowledged there are some very influential nations that have an alternate vision for our future. Well not the people actually, just the current elite variously described as the new aristocrats, the oligarchy, the Bielderbergers, the 100 or so controlling the bulk of the planets wealth etc. Let's call these influencers the 'elite'.

There is some evidence and a lot more rumour that this 'elite' has decided that international corporations, in particular, US/EU/Japanese corporations should run the world for the benefit of, well, the elite. That this elite has been dominating the western world's political and economic strategic actions since 1954, has manipulated and strengthened its influence, has bought and controlled the US government, subverted allied nations, facilitated one sided trade agreements, instigated conflict (black ops, regime change and war) to achieve its aims and has basically caused poverty in its unstoppable appetite for profit.

It, the elite, has further been accused for exploiting the people of their own states of origin by callously transferring manufacturing jobs to any cheap globally available sweatshop whilst crying crocodile tears into their dollar soaked handkerchiefs and moving motions at any global forum highlighting human rights and the evils of child labour. We the ignorant and frankly incredibly stupid people have unquestioningly accepted their proclamations that 'they didn't know and couldn't possible have known that there was child labour and poor working conditions'. I mean how could the elite that can run sophisticated intelligence operations against friend and ally know such commonly acknowledged facts?

Things were finally working out for the elite with the demise of the Soviet Union. This 'elite' tripped over its self in rushing to the now shrunken boarders of that old enemy Russia. Calling on its military resources (that's NATO in Europe), manufacturing threats, engineering the overthrow of Russian friendly leadership in Ukraine, trying to deny the oil and gas rich regions of the Middle East, hurtling headlong to create the nightmare that was a stable middle east, relegating Iraq, Libya, Syria etc. to the stone age. Lots of meddling, lots of helpful interference, lots of preventative action (denying Russia access to this and that resource) and extractions from theatres like Afghanistan (where it had done so much good) just to go and spread the help to other nations; times were good.

The 'elite' was also enjoying great financial success. Having suckered its own citizens into the sub-prime free for all and sharing the resulting financial crisis with the rest of the world they were secure that they would be rewarded by bailing each other out; they did control all the decision makers so – all good. They didn't mind that the costs of their unilateral and carefree exploits were charged to the 'maxed out' US credit card and that the people of America would assign their national mortgage to their next generations; they were secure in the knowledge that their global domination, their ability to print more dollars that all the world craved would set things right.

Then the elite looked over their shoulders and saw the new world. The first reaction was undoubtable outrage. How dare Russia and China cooperate? First they resist our UN domination (several incidents of Sino-Soviet vetoes of regime change strategies apparently weren't obvious enough). Then, without asking the elite, they announce the Silk Road Economic Belt and the 21st Century Maritime Silk Road, well it all got too much. Better stir up the China Sea issues and quickly tie up the rest of the world in trade pacts that preserve their dominance.

But how could the elite protect their privileged position. Easy, we'll continue our supremacist strategies and develop the **Investor State Dispute Resolution (ISDR)** system (ISDS). This system was first used in 1959 in a bilateral agreement between Germany and Pakistan to protect investors from discrimination and expropriation, refined in treaties such as the North America Free Trade Agreement (NAFTA – described by US Senator E. Warren as a trade pact model that that has spurred massive US trade deficits and job loss, downward pressure on wages, unprecedented levels of inequality and new floods of agricultural imports"). The elite must have concluded that 'that's the very one for us; lets run with that and make our global lackeys sign up to that right away'. They didn't even worry about the US constitutional legalities (the supermajority requirements; no need - they had that covered, our man will fast track that through).

The ISDS is well explained by the Attorney Ellen Brown[26]:

> "In The Economist, ISDS gives foreign firms a special right to apply to a secretive [offshore] tribunal of highly paid corporate lawyers for compensation whenever the government passes a law to do things that hurt corporate profits such things as discouraging smoking, protecting the environment or preventing a nuclear catastrophe."

She goes on to explain that the **Trade in Services Agreement (TiSA)** is

> "…about freeing international corporations from the government regulation necessary to protect the economy, the people, and the environment. They are about preserving privatized monopolies and preventing competition from the public sector. And they are about moving litigation offshore into private arbitrary tribunals."

The **Trans Pacific Pact (TPP** - Australia, Brunei, Canada, Chile, Japan, Malaysia, Mexico, New Zealand, Peru, Singapore, the United States and Vietnam) and the **Transatlantic Trade and Investment Partnership (TTIP** – Europe) will involve nations that collectively account for eighty percent of the world's Gross Domestic Product (GDP).

There are some very disturbing aspects of this 'treaty rich' world. The UN called for the suspension of US/EU negotiations on the TTIP because "we don't want a dystopian future in which corporations and not democratically elected governments call the shots"[27].

If we now look back over the organisations and institutions mentioned above we might be justified in expecting some good outcomes. The two most powerful global financiers both with the mission to alleviate poverty, global trade treaties, agreements, and groups of 'important nations' all committed to fair trade, all engaged in making the world fair and democratic and working for the good of ...? Thankfully we have a global organisation to at least regulate global trade.

The **World Trade Organisation** (**WTO**) was established to regulate international trade and promote free trade to stimulate growth. Just like the IMF and the World Bank it continues to be severely criticised. Its practices and trade policies are seen as spreading inequality through unrepresentative and non-inclusive decision making which denies the developing nations an effective voice. Little wonder that much of the world is rethinking the benefits of yet another old world institution.

It's not unreasonable to conclude that the global organisations that are regulating international trade and solving issues aren't actually helping much. Why is there such disparity in wealth, such inequality? Why are so many still starving? Why is poverty so wide spread? Where have all the massive profits gone if not to alleviate these problems?

The Four Kingdoms of the Third Millennium

There are of-course more than four kingdoms (the UN lists 206 states of which 193 are UN member states) but for overview purposes I've defined 'kingdoms' as a grouping of nations that are most likely to dominate the present and near future geo-political landscape as follows:

- The Failing Kingdom (US and Allies)
- The Future Kingdom (China, Russia, India, Brazil and South Africa)
- The Crumbling Kingdom (European Union)
- The Stateless Kingdom (Mega-Corporations, Terrorists and Crimes Incorporated)

I've not assigned every nation – that's too difficult for me but if you want to investigate current international influence on the African Continent, South America and likely alliances in the troubled Middle East you might well determine that China is actually assisting a great number of nations and doing this without political interference and... well, they're actually welcomed and appreciated. Anyway let's have a closer look at these provocatively titled kingdoms.

The Failing Kingdom

This is the kingdom of America and its allies; a huge kingdom of all those countries that align by choice or coercion with the American hegemony. It is the self-arrogated coalition, dominated by the American superpower that still believes that it is its right to dominate the world and allow other nations all the freedoms they desire as long as these don't interfere with its own strategic and commercial interests. It is a superpower that has been in decline for several decades and in denial for even longer.

Most of us have an image of America as a wealthy nation with a McDonald's or a Starbucks on every corner. Very few of us however know that there are more 'pay day lender' (short term, high interest loan provider) shop fronts in the US than all the McDonalds and Starbuck outlets combined. How is this possible and why might that be?

The cold war ended some 25 years ago. Since then the US has assumed the role of global hegemon. It has awarded itself the right to dominate the globe from Washington and Wall Street and has continued its tradition of dictating, punishing, rewarding, overthrowing non-compliant regimes, conducting covert destabilising operations, engaging in unlawful warfare, bullying and engaged in massive intelligence gathering on friend or foe alike. It has been doing this with an attitude of complete entitlement disguised as global peacemaker and all-round good guy. For the most part we've swallowed it.

I have many friends and family in this kingdom and nobody should derive any pleasure from acknowledging that we are collectively failing to lead and manage effectively. All of us that have been in leadership roles are to blame. There is however no way to fix anything if we lack the courage to admit we've made mistakes. Let's be honest, analyse our collective failures, forget the blame game and see if we, with the help of our thought leaders, can develop strategies, engage with the people and exploit technologies to get it right for the common good.

I absolutely stress that it is not the people of America and its allies that are to blame. They are victims, just like all of us, to sophisticated deception and manipulation. The US is not the utopian democracy it proclaims. Power is not wielded by its people nor is it wielded by its representative political leaders; it is exercised by the elite, a modern day international aristocracy, the 100 or so world's richest and influential people and the same 100 that have between them as much wealth as half the globe's population.

The good people of this Kingdom are not only misinformed and deliberately mislead, they are also suffering the benefits of runaway corporate and individual greed, greed that has seen record unemployment, radical reductions to living standards, some of the poorest healthcare in the world; the many good and honest people that have been sacrificing their children to conflicts around the world - spreading the good news of the democratic way. It really isn't the people; it's their leadership which in itself is often manipulated by the elite and powerful.

The US has racked up foreign debt (over US$18.5 trillion in total external debt as at June 2015) to the extent that the only acceptable truth its government can utter is "we will never have the money to pay you back". To date, that admission has been avoidable only because the above mentioned elite control the actions of the World Bank and the International Monetary Fund. You might, for yourself, research the US move from the liberation of its dollar from gold reserves, the creation of the petro-dollar to the liberal printing of monopoly dollars called quantitative easing. The recent revelation and penalties awarded for international exchange rate fixing by major world banks is also indicative of liberal practices that have for decades favoured those in the know.

The fact that American diplomats are still proclaiming the virtues of their hallowed democratic values and economic management to anyone still mildly interested would be laughable if it weren't so detrimental to its own population. If you think that's a bit harsh then consider these facts:

- Governance: There is general awareness among citizens of the US (and globally) that the deciding factor of government is money. "Ordinary citizens feel that their supposedly democratic government no longer truly reflects their interests and is under the control of a variety of shadowy elites[28]."

- In the US being in 'Government' does not mean being in 'Power'. Power is in the hands of economic interests – interests that mostly don't answer to the people of America; there appears to be no room for 'demos' (common people) in the deteriorated form so loosely described as democracy.

- Unemployment: Officially reported at 5.5 percent in May 2015 and unofficially argued that it is more like 22.9 percent[29]. But consider these facts:
  - 14.6 percent of the American population, that's 46.5 million out of 318.9 million people are on 'food stamps'. These food stamps are federal food purchasing aid to low and no-income people of the US.
  - There are about 149 million employed US citizens and 102 million non-working Americans (includes children). Of the 149 million, 26 million are part-time workers and some 9 million are self-employed. Does the ratio of 149 million working to 102 million not working look like a 5.5 percent unemployment rate?

- Infrastructure: The 2013 American Infrastructure Report Card issued by the American Society of Civil Engineers is a sad reflection of the real state of the union. In a grading system where A is exceptional, B is good, C is mediocre, D is poor and E is failing, America scored a D+. Not only is that not good, with an estimated required expenditure of 3.6 trillion dollars needed by 2020 improvements are clearly not likely. This once great superpower now ranks 16th in the world in infrastructure quality.

- Media: The US media is becoming world renowned for its high levels of inaccuracies, sensationalism, partisan positioning, poor coverage of important issues and globally myopic views. Ninety percent of America's media is controlled by a little over 200 media executives from GE, News-Corp, Disney, Viacom, Time Warner and CBS. Everything that the average American reads, watches, listens to and 'learns' and hence US public opinion, is controlled by a handful of media giants.

- Health Care: The abstract of the 2014 update report 'Mirror, Mirror on the Wall – How the Performance of the US Health Care System Compares Internationally' states "The United States health care system is the most expensive in the world, but comparative analyses consistently show the U.S. underperforms relative to other countries on most dimensions of performance. Among the 11 nations studied in this report—Australia, Canada, France, Germany, the Netherlands, New Zealand, Norway, Sweden, Switzerland, the United Kingdom, and the United States—the U.S. ranks last ..."[30]. Not surprising then that the US has the highest infant and maternal mortality rate and lower life expectancy than comparable nations. Information abounds on the approximately 50 million Americans that don't have health insurance, the 700,000 or so people that go into bankruptcy each year trying to meet their health care costs and the 45,000 people that die each year because they cannot afford health care. These are a few of the alarming facts that this superpower's market based health insurance system has brought about. Enough said really; if a country can't meet the health care needs of its population – what good is it?

- Education: 50 percent of all American college entrants do not graduate. Impressive statistic particularly when it is also acknowledged that the majority of those that do graduate do not possess the communication and computational skills that are needed to succeed in today's world. Not a stretch to conclude that they will fare badly in the more technical and complex world now emerging.

- Crime: "At some point, we as a country will have to reckon with the fact that this type of mass violence does not happen in other advanced countries". The sad admission by the US President Barack Obama following the June 2015 mass shooting of nine people inside a church in Charleston, South Carolina.

- Poverty: The United States Census Bureau reported the 2013 official poverty rate to be 14.5 percent[31]. That's 45.3 million Americans in poverty. Visit any American city and you will see clear evidence of this poverty; homeless people are everywhere. The answer that is being implemented by many city administrations is often to prohibit the sitting or lying around public spaces and sleeping in vehicles. Some have even passed legislation banning or restricting organisations from sharing food with the homeless in public places. No solutions to this unacceptable situation appear near; it seems that the US is simply trying to hide the problem. Shame.

- Inequality: This much publicised outcome of capitalism in action is really a logical outcome. Even Karl Marx had described this as a fundamental characteristic of capitalism when he wrote "Accumulation of wealth at one pole is at the same time accumulation of misery, agony of toil, slavery, ignorance, brutality, mental degradation at the opposite pole". Inequality in America is now at its highest level ever. One family (the Waltons of Walmart) have as much wealth as the poorest 42 percent of Americans and CEOs of large corporations earn as much as 300 times the salary of the average worker. Now, Walmart is actually closing stores around the nation – not a healthy sign or has the population shrunk that stores are no longer required? This obscene accumulation of wealth by the elite of America can only end in tears; little wonder that some of the very wealthy are already talking about and fearing the pitchfork wielding masses.

- Human Rights: Perhaps one of the more embarrassing characteristics of the USA today. This champion of democracy and self-appointed definer and champion of human rights arrogantly publish its annual State Department 'Human Rights Report'. The report makes comment and judgements on the human rights situation in many countries of the world but does not report on its own sad human rights record. It is well worth reading China's report 'Human Rights Record of the United States in 2014'. This report exposes America's supreme hypocrisy and details a well referenced and evidenced list of US human rights violations both in America and around the globe. From the spreading of gun violence, torture, racial discrimination, the injustices in civil, social and economic rights, police violence, spying and eavesdropping on citizens and foreigners, to America's violations of human rights in other countries; it's all spelled out in detail. It's not comfortable reading but it's necessary to understand the real picture.

All of the above is verifiable and is well understood by foreign nations and their leaders. They, after all, are interested in facts and not limited to information from partisan, profit motivated media that has lost any incentive to be factual or competent. I believe that the policy of catering to the lowest common denominator and the dumbing down of content to achieve that aim is not even contested.

I'm always surprised how senior foreign leaders can be so restrained and not burst out laughing when western leaders visit and start demanding that they need to this or that in order to come up to the 'western standard'. How stupid does America think the rest of the world is if they believe that anyone would want be create the nightmare that they are in.

The aims of liberal, representative or direct democracy as benign and effective systems of governance are, as I've said repeatedly, noble and attractive. The take away here is that before anyone seeks to hold democracy up as the only system, they really should implement it in reality first and demonstrate that it is sustainable in a capitalist regime. That argument has not been made.

This failing kingdom unfortunately persists in interfering around the globe. Still seeing itself as the supreme arbitrator on everything it has

LEADING INTO THE FUTURE

manipulated its European vassals to impose crippling economic sanctions against Russia. This however is not working as Russia is now re-establishing its local industries replacing past European and US import dependencies and establishing new and alternate trading partnerships. The result has been significant harm to the European economy. The annexation of former Soviet states (Eastern Europe, Balkans and Baltic States) and the encirclement of Russia with military bases are equally not providing progress and riches to the newly subordinated nations.

America's strategy to curtail China's rise in global dominance is failing. Having sought to surround China through expanded military base arrangements in Japan, the Philippines and Australia and establishing trade agreements with all and sundry (except China) it has failed to contain China. China and its now close ally, Russia, have taken the high ground. Not only has this new block deepened its trade and investment ties with major regional economies, as exemplified in BRICS, it is offering alternative banking which finally offers the world alternative financial support to the US dominated IMF and WB.

As the US continues to drive the deteriorating US-Japanese-European block and relying on military power and expansion to influence the globe, the new and emerging power, the Sino-Soviet alliance is using trade and economic cooperation to greater effect. The outcome is not certain. It appears that the US cannot step away from its old habits of creating and nurturing ethnic conflicts, military invasions, economic sanctions, self-serving trade agreements and regime change. How this all plays out in the near future will be interesting and could range from internal violent revolution to all-out war.

I'm not exaggerating here. How long will the massively growing dissatisfied citizens of the US put up with their deteriorating situation? Will a wounded and cornered past superpower simply whimper away or will they pursue a war of rectification; one where they will at once erase their foreign debt and destroy any challengers to their superior position. They may recall the bonanza that the Second World War was. The bonanza that had America producing half of the world's economic output and heralded three decades of prosperity and growth. Will they seek to replicate that and by default bring the global human experiment to an end?

The tone and observations in the above paragraphs reflect the content of speeches and press releases of presidential candidates in the 2016 US elections. I've also noted remarks and advice from visiting US citizens that have exercised their freedom of speech on Australian national television and implored us not to let our politicians make us like America and stressing "believe me - you don't want to be like us".

Referring once again to my Singularity University experience – we did an exercise where we sorted possible future news headlines into what we thought to be appropriate time frames. The headline "Inequality Riots in North America has China Call for Regime Change in US" was predicted for 2021; the audience was quiet when this was read out!

One point I need to clarify – the above is simply factual, most of the world knows this even if it is usually whispered. It is not my intention to blame America for all global ills – we all share that accomplishment and we need to remember that global instabilities have existed since time began; the middle east and the rest of the old world saw the rise and fall of many empires, waring empires that existed and disappeared long before the discovery of the Americas – so...

The Future Kingdom

A good understanding of what the Chinese government, and hence China, is focusing on can be gained by reading Premier Li Keqiang's report on the 'Work of the Government' which he delivered to the Twelfth National People's Congress in March 2015. The report details the wise coordination of national capacities to address their unique challenges, the real and substantive support to technology enterprises and frankly, evidences a pragmatic deliberateness that has become a defining characteristic of this superpower.

Faced with a demographic encompassing the full spectrum from rurally impoverished communities to some of the most modern cities and societies in the world supporting very technically advanced institutions, China is uniquely poised to be the dominant kingdom of the future.

Not a prediction but a simple logical conclusion that has precedence. Just as Germany and Japan were forced to rethink, retool and rebuild from the ravages of the Second World. They had no alternative but to do this with the then current technology; effectively building modern manufacturing based economic and social advantages that are still evident today. Given China's massive requirement to 'modernise' nationally, its political imperative to progressively share abundance and the managed application of exponential technologies – what's your conclusion?

Without resorting to political debate or a discussion of idealism there are some hugely consequential differences in how technology is driven and managed between the traditional 'western democratic world' and the Chinese world of 'Socialism with Chinese Characteristics'. The western model appears quite straight forward: principally private enterprise, some private public partnerships, basically the 'free' market decides what to develop, exploit and monetise and ever more vulnerable governments regulate reactively. In China, the government, one not subject to election cycle policy making, sets the agenda, encourages and often creates tech giants and manages the resulting technology.

There is still a persistent, if reducing practice, of underestimating China's technological achievements and there are several examples that do highlight significant challenges such as its networks' abilities to cope with big data. Whatever you may deduce there can be little doubt that they have taken huge steps towards technological advancement and created behemoths of research and development; all with ambient global reach.

The Crumbling Kingdom

The ongoing refugee crisis carries with it the prospect of countries putting permanent border controls back in place. Some of the 26 countries in the Schengen Area have already reintroduced short-term border controls. Austria, Germany, Denmark and Sweden are among those who have decided to reinstate controls.

Whilst indeed impeding the free flow of refugees these border controls also hamper the free flow of traffic and therefore slow down the cross border exchange of goods and services. As the refugee crisis shows no sign of abating and with no prospect of securing Europe's common outward border, re-establishing controls within could become a permanent fixture again.

The European Union has been described as "…economically and culturally torn, massively plagued by migration issues, ethnically selective, past conquerors, a permanent conference centre, and a union of unhappy and mostly bankrupt allies…" – surely not! From the Iberian Peninsula in the west, across stable Italy, stretching east to Greece and Turkey (not mentioning those troublesome ex- Soviet Bloc nations, bordered in the South by ex-crusader countries and including the satisfied Scott's in the far North - surely this kingdom has few concerns.

Other than the size of the EU's consumer market, some 500 million people, this kingdom has little to offer the new world. The developing world is rapidly negating any technological advantages that Europe once held and it is increasingly severing any dependence on this collection of old masters. In plain words – Europe, wake up, the world needs you far less then you now need it! If that's not a pleasant reality then remember also that future dealings with past 'vassal' states and ex colonial nations might extract apologies before dealing with you – alas on their terms now!

This Kingdom is also severely troubled by its relationship with the 'Failing Kingdom'. Still subservient to the Americans for saving them in past wars, still beholden to them for keeping the Russian Bear caged and slowly coming to the realisation that American interests are in the best interests of well – America, and specifically, the elite that is the real US. Recent regime change plans for Ukraine have dragged a dizzy Europe into sanctions that have severely hurt the EU and assisted Russia to refocus its internal economy. This is what Russia needed because it had developed into a retail economy (buying internationally to sell domestically) and had lost its ability to grow and manufacture for itself. The fact that isolating Russia is a stupid policy escaped these European subordinates who really should realise that if ever a nation could survive on its own devices – it's Russia!

Whilst I'm writing about Russia I have to admit that I do know why it took back the Crimea. Understanding the motives that created the Ukrainian conflict, Russia simply secured that real estate vital to its defence interests – and clearly the prize that US/EU 'assistance' to Ukraine was aimed at securing. Now that this desirable objective has been snatched from the West it's very likely that US/EU aid to Ukraine will gradually fade.

The joyful rush of the US/EU to 'grab, seduce and coerce' the ex-soviet nations to join the West was a very transparent grab to surround Russia, expand NATO and to gain new political and trade subordinate countries. Selling the 'American Dream' to these liberated nations, affordable through loans from charitable bankers like IMF, ECB and the World Bank, all the while denying the collapsing and unsustainable union of Europe. Whilst the few mid and northern European countries are the powerhouses of the EU, the rest, including Spain, Portugal, Iceland, Greece and Italy etc. will continue to find the union unpalatable. As austerity and poverty grows this kingdom will continue to crumble – and to quote T.S. Elliot "not with a bang but a whimper".

Europe also faces the very serious consequences of its, and the US's, meddling in the Middle East and North Africa. Their joint democratic reconstruction efforts in the Middle East, those efforts seeking to impose political regime changes, have been backfiring relentlessly and have resulted in unprecedented and unstable regional governance; the emergence of ISIS is just one of the outcomes; the displacement of millions of people fuelling the current refugee crisis is another.

The refugees flooding across the Mediterranean into Europe; some 220,000 in 2014, and now well over a million are a huge humanitarian crisis for the EU; a crisis, which the Union seems to have little interest in solving. This is demonstrated by the many countries refusing to take their share of refugees leaving nations like Italy to their own devices. This is also plunging Europe, nation by nation into an era of patriotic extremism. This is playing out on our media channels on a daily basis and does anyone see a humane resolution to this crisis and the worsening emergence of self-interest?

Surely this crumbling kingdom has so far failed to acknowledge that their support of interventions in the refugee generating regions has brought about a growing crisis; they have certainly failed to accept their responsibilities to assist and aid the swelling arrivals of 'illegal' refugees.

Agree or disagree, the Western Powers' attempts to alter the political landscape of the Middle East, has destabilised the entire region. Already drowning in the US/EU self-created financial crises these kingdoms have also managed to add a refugee crisis to the chaos that exists in Europe.

For this kingdom to still shriek the benefits of their values (peace, democracy, tolerance and freedom) to the rest of the world is not only sad and pathetic - it is a recipe for disasters and signs of conflicts yet to come. This 'Crumbling Kingdom', its master 'Failing Kingdom' and their entourage of lackey followers (unfortunately includes Australia and Canada) need to reflect on their self-righteous meddling – has it all been worth it – for whom?

Europe, its demonstrative failures and the potential for these to result in a catastrophic destruction of the union and many of its member states, may be exactly what the developed world needs to reform its dysfunctional global democracies. That's a big statement; I will return to this question in Part 3.

The Stateless Kingdom (Mega-Corporations, Terrorists and Crime Incorporated)

This kingdom is made up of all those organisations, groupings, collectives and individuals that are not subject to any national jurisdiction. It includes the major international corporations that can avoid taxation or restrictive compliances by having its feet in many countries including tax havens. These are the corporations that will benefit from TPP like trade treaties; those conglomerates whose commercial objectives will over-ride sovereign governance. They are too powerful to be controlled by any one state; they have become independent of national interests. It is these corporations that can at will step-outside any law, exploit governance weaknesses, buy influence, export jobs to whatever country provides the cheapest human labour; they are both the cause and effect of unrestrained capitalism.

These powerful corporations have been enabled by the accommodating policies of the 'West', policies of global financial structuring and manipulation that has collapsed the economies of countries and impoverished billions.

It is these mega-corporations that largely create and maintain the elite; those 100 or so most powerful people that are the actual global influencers. It is their will that the western world reacts too. As unbelievable as that might appear to be – if you research this (the hyper-connectedness of communication and social media will in any case reveal these realities more and more) you will discover who actually controls the US Government (and hence all lackey states), who shapes global policies, owns and controls western media, directs financial and military focus and to what self-serving purposes the democratisation and globalisation arguments are actually being used for.

Believe it or not – it doesn't change the facts, facts that are easily verifiable if you care to look. Note that I am not making any judgement on good or bad; the issue here is simply that before we collectively think about the future and any abundance for all, we really should understand what's been happening. Those that won't look, those that are too busy or too comfortable or those that just 'don't want to know' will indeed be surprised in the near future. Impoverished, dispossessed and left behind people will soon find their voice and much of the world will not like the tone and the changes they will bring.

So far, we've considered the known global context, the apparent, sometime legitimate overt world. The world today however also includes a huge shadowy world, the world of organised crime, black economies, rouge organisations (sometimes supported by governments often independent, black ops and intelligence organisations again sometimes directed by governments, corporations or individuals), terrorist organisations and individuals that pose a constant and growing threat to all.

This is the Kingdom that poses a clear and present danger to mankind and it's the kingdom that has at its disposal all the 'benefits of exponential technology' and that means – long before we, the trusting citizens, get any benefit from that same technology. Criminal organisations don't need approvals from anyone to implement and misuse the latest cyber-crime tools; to use sophisticated weapons (even weapons of mass destruction) against their targets. Extremist terrorist organisations are not limited in what they will use to achieve their goals; when their goals are to bring about Armageddon (i.e. ISIS) what might they be willing to do?

It is not my intention to detail here various global 'end game' scenarios but I need to acknowledge that our future will in all probability be severely tested by horrifying and devastating actions of organisations and terrorists.

I urge you read Marc Goodman's recent book 'Future Crimes'. Marc is 'super experienced' in law enforcement and the latest technology; the founder of the Future Crimes Institute and the Chair for Policy, Law and Ethics at Singularity University. I strongly recommend his book, but it's a hard read – full of fact based information that will in his words provide "A friendly warning: if you proceed in reading the pages that follow, you will never look at your car, smart phone, or vacuum cleaner the same way again". It will also alert you to what future crimes will be and expose the frightening tools of crime and terrorism that are already being prepared.

My reason for lifting the veil on uncomfortable truths, challenging accepted paradigms and frankly telling it 'as it is', is not to frighten anyone or to criticise our fore-fathers. It's simply to establish that leading into the future is our greatest human challenge; it will require global governance and it will require outstanding individuals to lead us. Above all else, it will require people like you and me to communicate, to understand, to support each other, to add value to ourselves and to the add value to the global community; a community that is in transition, a transition to what? - That's the challenge isn't it?

The World Today

However we analyse our world, regardless of our objectivity and perhaps entrenched subjective interpretations, we cannot deny the facts. Whatever we think of our global situation, the present context that is our starting point for the future, we should clearly understand the challenge. Existing technology could meet all of our basic human rights challenges today (water, food, sanitation, shelter and a peaceful existence for all) if we were to create and accept an effective global governance structure.

We often congratulate ourselves for lifting this or that many million people out of poverty. We should in many cases first re-examine what we have done to create that poverty in the first place.

We pat ourselves on the back for having liberated various countries from a variety of unpleasant rulers or governments, for having introduced our religions to replace their own hedonistic practices and of having exported our cultural values etc. sickening no?

The most ridiculous and stupid posturing that we are all too often guilty off is to force our inept application and misrepresentation of 'democracy' on other nations. There is nothing wrong with democracy – except that it doesn't work in any nation of the world. Yes, the democratic values are great but even Plato understood it better two and a half millennia ago than our citizens and leaders do today. Our labels 'socialism, capitalism, democracy etc. are unthinkingly thrown about as good or bad; really? That's the best we've come up with? How much more inequality, poverty, racism, supremacist attitudes and dumbing down of populations is it going to take to make us collectively realise that the western democracies are a myth. That these systems of government are not representative and are severely failing their people?

Here are some clues (much in the same way that a punch in the face is a hint of disagreement):

- 2.4 billion people worldwide still lack basic sanitation[32]
- The UN General Assembly's affirmation that "democracy is a universal value based on the freely expressed will of people to determine their political, economic, social and cultural systems and their full participation in all aspects of their lives," – note it does not qualify this by adding that the US can change those governments not to its liking
- At the end of 2014, there were 59.5 million refugees in the world[33]
- Oxfam reports that the wealthiest 1 percent of the global population will own more than the remaining 99 percent by 2016 and "the richest 80 individuals in the world had the same wealth as the poorest 50 percent of the entire population, some 3.5 billion people"[34].
- The world is dominated by two economic peaks – the twin peaks of deficits and surplus. Whilst the global deficit is huge, the surplus is even greater. The world economy is allowing these surpluses to flow to an ever shrinking number of people because no effective global surplus distribution system exists – that's simply not sustainable.

Really - we want a Global Democracy?

The above sections paint a somewhat grim picture of the globe today. Too many are in denial about historical fact and many more are still selling their version of the ideal.

The once all powerful 'western world' is still revelling in its belief that it has triumphed and demonstrated that its brand of democracy has defeated communist Russia and is self-congratulatory about the seductive embrace of capitalistic practices in that other worrying nation – China, would undoubtedly champion a global democracy. After all, that is what the US and its allies have been espousing for a long time; haven't they been trying to introduce democracy through economic, military and covert means for decades wherever they had influence?

Given the dysfunctional US and European pseudo - democracies described above, why would anyone subscribe to their interpretations? Notwithstanding this, assuming the West has already decided this question for us; let's consider just how a global democracy might work.

The Oxford dictionary defines it as "a system of government by the whole population, usu. through elected representatives." There is however no consensus on its definition even though some fundamental characteristics seem to be widely accepted. Larry Diamond[35] provides a useful set of characteristics of democracy:

- A political system for choosing and replacing the government through free and fair elections
- The active participation of the people, as citizens, in politics and civic life
- Protection of the human rights of all citizens
- A rule of law, in which the laws and procedures apply equally to all citizens

So, one person, one vote, globally, seems to be the right interpretation. How would that work for the current champions of democracy? Assuming one couldn't argue against proportional representation then the democratic global government's 'house of representatives' would (based on say one member per 10 million people and a population of 7.4 billion) have 740 members. The 20 most populous nations would have the following representatives:

- China - 140
- India - 128
- United States – 32
- Indonesia - 26
- Brazil - 20
- Pakistan - 19

- Nigeria - 18
- Bangladesh - 16
- Russia - 14
- Japan, Mexico - 13 (each)
- Philippines, Ethiopia - 10
- Vietnam - 9
- Egypt, Germany, Iran, Turkey – 8
- Congo, Thailand – 7

Assuming nations would be tempted to vote along 'party lines' how globally relevant might the desires and influences of our champions of democracy be? I mention this because we do need to understand the fundamental flaw of global interactions justified on the basis of a genuine desire to democratise the world. The aim sounds noble but its realisation would not please those nations well served by their current politics.

There are however a growing number of nations that have observed and studied existing democracies; again I'm not describing the theoretically ideal democracy, rather those misleading manifestations disguised as such. These countries don't want a democracy like the US, Europe and their faithful followers and allies (e.g. Australia, Canada and New Zealand)? These nations are now revealing examples of factual distortions and spectacular governance failures in countries like the US. The report 'Human Rights Record of the United States in 2014' (The State Council Information Office of the People's Republic of China, June 2015) is one example. Other government's failures, hypocrisies and inequities are also being exposed at alarming rates due to emerging information technologies, social media and data leaks.

Democratic values are however universally accepted as good and proper. One need only look at China which although conveniently labelled as communist, is embracing and enacting democratic principles. Their government is demonstrating a committed transition to implement their interpretation of a socially responsible democracy but they are doing it their way. They are demonstrating and exercising their right to develop a system of governance that works for them – as is the fundamental right of any sovereign nation.

Zhang Dejiang, the Chairman of the Standing Committee of China's National People's Congress, during a speech he gave in August 2015 at the Fourth World Conference of Speakers of Parliament at the United Nations called on countries to "seek peaceful settlement of differences and disputes through dialogue and consultation, promote democracy in international relations, and safeguard international equality and justice."

Isn't it interesting and somewhat odd that the labelled 'communist' China is actually the loudest and most convincing voice calling for democracy in international affairs? What are even more noteworthy are China's actions (always louder than words). It doesn't just promote values – it demonstrates them in action. No finer example of inclusive, respectful, sustainable and mutually beneficial work exists than the Chinese President's, Xi Jinping's 'Silk Road Economic Belt and 21st Century Maritime Silk Road' initiative. What comparable initiative has come from 'western' leaders? (Political interference, bombing and regime change doesn't count).

My observation is startlingly clear – the greatest force for a truly democratic world is coming from China and Russia. It is these two powers, thankfully emerging in influence, that have become the champions of sovereignty and global fairness. The self-serving nature of the western powers is shamefully being exposed and seen for what it has always been. I know that some of you will disagree with the above but if you look at the various ways that superpowers are influencing world affairs – then you might come to the same conclusion. We unfortunately continue to suffer from inherited differences and adopted beliefs that do not reflect reality. We need to ditch old misconceptions and we need to judge, evaluate and think afresh on the paradigms that we have conveniently accepted for far too long.

The Incorrectly Labelled World

It's very human to simplify issues and apply convenient labels. This non-heuristic trait allows us to accept judgements, opinions and meanings that are in line with our own beliefs and save us the trouble to discover or learn something for ourselves. Some of the convenient simplifications include:

Labels that we routinely apply to countries include 'developed' and 'developing', 'first' and 'third world' are arrogant and judgemental. We apply these as if the majority of western nations are finished; i.e. they're done, fully developed and perfect. They have reached the end of development, their utopia, they have progressed and achieved all – hence they are developed. Those unfortunates that are still developing are assessed against the values and ideals of the developed world. We proudly hold ourselves up as the model, the end product of development; the inherent righteousness of our systems and simply the way we do things are broadcast to all as global aspirational objectives. It might well be convenient to apply these labels but they are arrogant and insulting to the majority of the world; we are all developing nations; some are simply doing it their way and from their inherited starting point.

We are all born equal. Really – the baby born in sub-Saharan Africa to an AIDS infected mother needing to decide whether to feed her new born or to let it die so that its older siblings might have a better chance to survive is equal to the baby paraded on the royal balcony or to the one born to the wealthy couple that has already secured a place for their newborn at that exclusive pre-school. No, we are definitely not born equal but we are mostly born with a joy for life, a love for others, an insatiable curiosity, an appreciation of difference, a desire to live in the moment, happy, sharing, playful, kind etc. We are born – well just about perfect. Then we learn.

A democracy is the only proper way to govern a nation. This one is so limiting and unrealistic that it defies logic. Firstly, we can't agree on the definition or real implementation of democracy (direct or representational, its modern interpretation justified on an expedient 800 year old document – the Magna Carta, the unenforced separation of powers and the misrepresentation of the people by elected officials that all too often are bought or influenced by special interest groups). Secondly, the pitfalls of democratic political models were apparent to philosophers like Plato and

Aristotle more than two and a half thousand years ago. Thirdly there is no purely democratic government anywhere in the world.

The belief of entitlement and righteousness - from the constant abuse of the freedom of speech, that a belief system or religion is superior to others to 'God is on our side' or racial superiority, these applied labels are the most stupid.

The above are just a few 'conveniences' of our world. Agree or disagree; it won't alter the fact that if we, as a global and fair community, are to benefit from the incredible good that exponential technology can bring then we first need to understand and acknowledge the state of play. Only then, can we change what we jointly agree needs changing, only then, can we form a unity of purpose. Anything short of that will keep us spiralling into a divisive and predictably unpleasant future.

Environmental Change

We need to also acknowledge that our global context involves our globe – clearly! I'm not going to delve into cosmic environmental events. Whilst there is a real possibility that the Earth could be struck by another meteor, like the one that sealed it for the dinosaurs, I'm reasonably certain that a 'Will Smith' (of 'Independence Day' fame) will most likely save us all anyway.

But there are some natural catastrophes that we should give some thought to. What if we had another volcanic eruption the scale of Krakatoa? An eruption that might consign the earth to a decade of sun-less winter, that disabled our satellites? What if the tectonic plates moved enough to 'sink' large portions of continents? How ready are we to jointly deal with what nature might well throw our way?

Change can come from unexpected and existential directions; some we can do little about, others we should consider. It is the man-made effects, the human impacts that threaten our precious environment that we need to focus on – that's all we can do at this stage. Climate change, the

119

evidenced warming of our planet, has been accepted as one such threat, one acknowledged by the UN and the reason for our current global focus on Sustainable Development Goals

The Earth, so apparently solid and permanent is nonetheless fragile. Think of the very tiny layer, that layer defined by the Earth's thin crust and appreciate that we humans can exist only in a thin layer. Our seemingly large planet with a mean radius of 6,371 km and an equatorial circumference of 40,075 km nevertheless only affords us a small spherical layer of about 10 km in which we can exist. Go too deep, gravity will crush you, go too high – not enough oxygen. Of the globe, only 0.47 percent is habitable by human beings; so 99.53 percent of what we see as our planet is off limits to us. It would seem quite smart to look after that precious thin layer.

The environment, that very finely tuned ecosystem, is being severely strained by our human occupation. The obvious culprits, depletion of resources such as oil and gas, the air and water pollutants, industrial processes that alter the very makeup of our atmospheres are well known to be inflicting real change. Regardless of the UN's clear position and acknowledgement of the threat that global warming poses debate still rages on and so do our environmentally damaging activities.

Closer to Home

Where I live, in Australia, there is little leadership, poor management and no vision, politically, socially and in too many boardrooms. I've witnessed astonishingly directionless, disorganised and grossly incompetent middle to senior management in all levels of government. This strong statement isn't even challenged by the very people working or living here! Agree? Don't agree? What am I'm missing?

Perhaps the above observations are just localised problems. What's been your experience and what is your awareness of what is really happening in the halls of government departments and major corporations?

Are you happy with your government's vision for the future and their ability to manage its implementation? Satisfied that your national and regional leaders were selected on the basis of demonstrated knowledge and leadership ability, the ability to leverage the most capable for the good of the nation? Do you perceive that your government leaders are working to achieve the objects that are prescribed by the population (following a particularly detailed and inspirational selection process) rather than governing for an election cycle? Do your leaders promise the impossible or do these leaders communicate doubts and challenges but still reassure you with hope and trust?

What is our national higher purpose? What are the nation's declared values and how do they account for your values?

Clearly, the above are generalisations; of course good leadership exists. We do occasionally find visionary leaders; there have been quite a few in the past and we also have people who are exceptional managers. Unfortunately, based on my experience, these are exceptions.

Any, even cursory, overview of our particular corner of the world today shows us that there is an increasingly urgent need for change.

As I've already asserted, globally we are not well governed; rather the world resembles a sandbox filled with difficult children barely tolerating each other and obsessed with a perceived necessity to focus on self-interest. Don't agree – travel to foreign nations, experience the warmth at immigration, the personalised scanning, padding and dog sniffing coupled with the begrudging processing of your passport – perhaps not signs of brotherly love and trust but a good indication that all is not so well.

Major corporations have grown in influence and political power to the extent that their desires over-ride even pretend democratic processes, bypass taxes and simply concentrate wealth to an ever shrinking minority. There has been much recent media hysteria about the fact that the richest of the world's population owns the majority of wealth. Not a new concept really! When was the phrase 'the rich are getting richer and the poor are getting poorer' first coined and what has been done about that if it was perceived as an issue?

Much of the world has adopted systems of education, in those lucky nations where education for most is at least possible, that stifle innovation, creativity, hope and most alarmingly, is significantly lacking in relevance to today's fast paced global environment.

In an increasing number of countries the young are made to pay for education; by this abhorrent practice we turn our back on our most fundamental responsibilities of teaching, developing and enabling our young to reach their potential. Few animals abrogate the responsibility to empower their offspring; sadly, we humans do all too often make others responsible. Many 'modern' nations now make the child responsible for their own development; we widely acknowledge that education has been failing our young for some time but we have not fixed the problem.

Gladly there are nations such as Finland; clearly a smart nation; emerging as leaders in introducing super progressive educational reforms. South Korea, often criticised by soft observers for the high pressure, high performance and associated social issues that their disciplined approach requires, has perhaps a very sound strategy [survival of the fittest] to enable their young to cope and thrive in the hyper competitive challenging near future.

Global health care, doesn't actually exist, its mostly limited to wealthy countries and within those countries, to those that can pay the most, and to those fortunate enough to be occasionally indulged through charitable people and temporary, if reluctant foreign aid from donor countries. Where such aid is given, it is more often a form of tokenism; a reaction to what filled the television screens in lounges of the first world and was needed in order to appease the suddenly but only temporarily caring world. Whilst I absolutely acknowledge the exceptional work done by philanthropists and of many charitable and volunteer organisations around the world. Isn't that in itself a real worry? Why is it that they are needed? How good are legitimate national governments and how are the many global governing bodies doing when charity is required to assist the neediest?

Beliefs and spiritual values are becoming ever more divisive in an apparent attempt to dominate collective minds. The western world holds steadfastly to the belief that they are right. God is obviously on their side and has been ever since the holy crusades devastated Europe's southern neighbours - North Africa and the Middle East. That, coupled with the rape and pillage sometimes described as colonialism, might just explain the degree of "push-back" that is growing globally both in severity and extremism.

So What?

Change is inevitable and necessary. Our common history, differently interpreted perhaps, has laid the foundation and we cannot build a future based on any false foundation. Everything is as it should be; whatever is good or bad in our world has been caused and manifested. Whether your view is a realised utopian world or a more realistic conglomeration of problems, one thing is certain – it's time to accept responsibility and to manage inevitable change. Life on earth has continuously been changing our society and environment and we are about to experience a much accelerated, technology enabled, rate of change that will challenge all of us. Exponential technology, whilst it can be applied to create abundance for all is already upon us. Today, whilst voices demanding good global governance are getting louder, this technology is already effecting and shaping the global community. It is heralding change, revolutionary change and most national governments are simply not engaged.

Our internet enabled hyper-connectedness coupled with the trillion soon to be connected sensors (the internet of things) has been described as the nervous system of the globe. That does seem apt, everything and just about everyone connected, communicating and sharing. There is however a problem with that and comparing it to the human nervous system is unfortunate; the human nervous system has a brain. This control centre is of course vital; no advanced nervous system can function without one. So if we are to liken the future internet to a human nervous system then what is or will control it? Hail ~~Google~~ Baidu in the cloud!

I believe that the current world's nervous system is suffering. You could make up a global 'renaming map' (breaking down the before mentioned four kingdoms) allocating diseases like self-harming, dementia, selective amnesia, bipolar disorder, apathy, depression, anxiety, stress, schizophrenia and the very popular attention deficit disorder to various nations (multiple disease allocation is allowed) and you could then sprinkle that map with some other ailments. Freely and accurately add: starvation, obesity, racism, superiority complex, entitlement, greed, kind, displaced, redundant, irrelevant, war-torn, migratory, closed, and don't forget other labels like under-developed, 1st and 3rd world etc.

Whilst the above might appear frivolous, it's not. Think about national conduct and international perceptions; why we might label a country or assign some characteristic. Are we expressing our knowledge, our understanding, our prejudices and how real are our perceptions? Our world needs a social reality and morality check.

Given the present global disconnect, the lack of any form of unified governance, no 'brain' or clearly defined future and add to that environment the very imminent impacts of exponential technology and it should be no surprise that a massively growing number of individuals are finding it hard to adapt and cope. We may be able to manage that global nervous system, our world, with a single brain, or perhaps multiple brains, but intelligent governance, leadership, is desperately needed. Our future depends on it.

PART 2

THE TECHNOLOGY

CHAPTER 6

Exponential Technology

"It is our happiness to live in one of those eventful periods of intellectual and moral history, when the oft-closed gates of discovery and reform stand open at their widest. How long these good days may last, we cannot tell. It may be that the increasing power and range of the scientific method, with its stringency of argument and constant check of fact, may start the world on a more steady and continuous course of progress than it has moved on heretofore." (Edward Burnett Taylor)*

Why it Matters

In the early chapters I wrote about my 'waking up' to what's happening with technology. My alarm wasn't just because I felt that I had been left behind. No, what was life changing was the realisation that the technological impacts were real, evidenced and unstoppable.

The realisation that really hit home was that humanity is on the path to transcend the limitations that we have in the past accepted as defining human characteristics; limitations accepted as the end product of the biological evolution that has evolved modern man.

I expect that if the above doesn't sound convincing to you at this stage that you are in for a surprise when you read and think on the implications of the technologies detailed in the following pages. What do you think the near future human; say within 25 years, will be and look like? Would you think it possible for you to live as long as you choose? Do you accept that you could be embodied in a body of your choice; a fit body at its peak; one free of disease; one able to self-repair and survive in extreme environments? Would you believe that your intelligence, your memory and thinking abilities could be greater than all the presently alive human brains put together?

Perhaps not – but there's more; would you accept that humanity will have the ability to create any current, new and edited biological lifeform and any physical substance or material from readily available resources? That humanity will be able to manufacture anything from fundamental atomic scale building blocks?

It all sounds unreal and like hyped-up science fiction doesn't it? Well the fact is that it is not. It's real, it's currently under development and its implications are massive. Our near future is being created now and its global impacts will challenge all that we have believed about our world and redefine what is possible.

This book is written to help you to see and understand this clearly. None of us can predict the future; all of us can however learn what's going on and we can anticipate outcomes and acknowledge what science, human ingenuity and applied intelligence is creating – what is 'leading us into the future'.

One of the greatest minds of our time is a man named Ray Kurzweil. He is a Co-Founder (with Peter Diamandis) and Chancellor of Singularity University; his published biography[36] (http://singularityu.org/bio/ray-kurzweil/) is worth noting.

He holds twenty honorary Doctorates and I trust you agree that he's probably quite smart, a serial achiever, and a thinker of consequence whose deliberations deserve serious consideration. I have studied his 2005 book

"The Singularity is Near – when Humans Transcend Biology" and have recognised it as the foundation to much that has been written, reproduced and debated ever since. I'm no different and I've leant heavily on his work in the pages to come.

Whatever your take is on the state of the global nation, you probably know and feel that all is not well. You might conclude that we are in for either a better or worse time but I hope that you recognise that we are in for significant changes. You might choose not to know, to accept whatever the future brings and find a practical 'can't do much about it' position somewhat comfortable. Clearly that's OK. If on the other hand you feel the need to be informed, to participate and understand the opportunities and threats that we will, individually and collectively, face then do critically consider exponential technologies and all that it means.

The publication of 'The Singularity is Near' was ground breaking and remains a definitive logic study of science not because it's the only treatise of a technology shaped future but precisely because it was written by Ray Kurzweil. He is the man on the spot; in the know and part of the collegiate of thought leaders that are making it happen. This is no armchair 'what might be' expose or a new series script for a 'Space Odyssey' sequel. It is an explanation of why it is so and why it matters, and it is still today, discussed and debated; it remains highly relevant and a verifiable roadmap of our progress towards actualisation of technological progress.

Not surprisingly given the increasing attention that technology is getting a lot is being said written and speculated about technology, its exponential rate of development, and the disruption to the status quo, its convergence and the many benefits and inherent risks this will bring to mankind. There is little doubt that the era of rapid innovation across all disciplines of human interest is upon us and is indeed occurring at an ever increasing rate. We are increasingly being exposed to numerous sound and image bytes that promise everything from immortality and instant cures to technology that will enable us to access a cloud based neural cortex to increase our thinking powers a thousand fold when needed.

There have been some other insightfully revealing books published recently on exponential technology and the one that set the scene is 'Abundance: The Future Is Better than You Think' by Peter H. Diamandis and Steven Kotler. This positive, scientifically justified representation of very probable technological developments and their impacts on humankind and our planet, advocates that abundance for all is not only possible but details the technologies that can make this happen. The book reveals how four key factors, exponential technologies, the innovator, techno-philanthropists and the rising billions (a reference to the expected billions of additional internet connected people) can combine to make abundance for all achievable.

Diamandis and Kotler published 'Bold: How to Go Big, Create Wealth and Impact the World' in 2015. A sequel to 'Abundance', 'Bold' is an experienced based blueprint on how leaders (exponential entrepreneurs) might position themselves and apply exponential and disruptive technologies to address global challenges, to tap into crowd based funding, sourcing and incentives in today's hyper-connected world. As President Clinton has said - "Bold is a visionary roadmap for people who believe they can change the world – and offers invaluable advice about bringing together the partners and technologies to help them do it."

It's no surprise that the implications of exponential and disruptive technologies would be investigated against today's global corporate environment. Salim Ismail, Michael S. Malone and Yuri van Geest did just that in their collaborative work 'Exponential Organizations: Why New Organizations Are Ten Times Better, Faster, Cheaper Than Yours (and What to Do About It)'. They detail their understanding of the characteristics of exponential organisations that leverage on big data, algorithms, community, the crowd and exponential technologies to radically outperform conventional organisations. With numerous examples of enterprises adopting various levels of exponential organisation the book even describes the roles and responsibility changes required by Chief Executives.

In the words of Ismail: "Exponential organizations are the future of commerce, non-profits and even government. It is the only model that can keep up..."

Exponential technology, its duality, the ability to assist and to disrupt, whilst tackling our greatest challenges is not just being seriously analysed, studied and debated globally – it's actually being realised. Too many of us however are largely unaware that the exponential technology revolution has begun. Limited to the occasional exposure to the newsworthy, most people are simply not aware of the impending changes such technology will impose and have not yet acknowledged that early consequences, like the shrinking job market and the devaluation of traditional education are manifestations of technological progress already applied.

The publications mentioned above provide compelling insights into why technology matters and really 'should matter a great deal' to all of us. Technology once developed becomes part of our human evolutionary experience. Such technology will be used and developed further. I chose the above mentioned texts because they explain what is possible and how exponential technologies are likely to affect us. They are admittedly 'Silicon Valley' centric and reflect the regional adoption of the 'exponentials' argument but they are ground-breaking and definitive works that really ought to be read by managers and leaders everywhere. Whilst many executives (Fortune 500 C-level, government officials, managers and leaders generally) are being briefed at an urgent pace at SU and similar global hubs, most can't take advantage of such briefings. This is one of the reasons why I'm writing this book. An overview, a reality check and to address the 'so what' and 'how will this affect me?'

If you want more evidence of present and imminent exponential technology and innovation driven effects on business you could read Deloitte's sixth Technology Trends Report[37] or any number of similar reports that leaders are describing as 'must know'.

Linear and Exponential

In Chapter 2, The Human Condition, I described the human brain and how its performance compares with computers. Regardless of our human brain's massively parallel processing capacity that for now persists to outperform computers in pattern recognition tasks, we are nevertheless linear thinkers. We perceive our environment, our experiences in time (and of change in general), in a sequential moment to moment fashion. Our minds have simply developed in that way and it allows us to comprehend and differentiate past, present and future events. We think in understandable steps ahead, much in the way that we might consider a possible chess game move, one, two or several steps ahead of actual play. We imagine these steps as constant sized increments similar to time ahead; in hours, days, months and so on. Each step forward is a standard addition of a time period that we have learned to trust because it reflects our observations and experiences.

I made the point that our V1 brain was developed some 200,000 years ago and has had no significant hardware or software upgrade since then. We are however taxing the software with our modern complex environments; a lot to learn, understand and make sense of. Our perception of time and change is housed in that same V1 model brain. We are preconditioned to perceive and assimilate change at the pace of change that existed when the brain evolved its then design characteristics. This human propensity is often described as linear thinking and in numerous explanations of exponential technology concepts we are invited to consider the 'human steps' example.

Here we are asked to imagine taking thirty linear steps, regular normal human walking steps. It is then revealed that that would put us – wait for it – thirty steps, say 30 meters away! Then we are asked to imagine taking 30 exponential steps where exponential means that each new step is twice as long as the previous step. That means our steps would be 1, then 2, then 4, then 8, 16, 32, 64, 128, 265 … and so meters long. Our last step, our thirtieth step, would be 536,870,912 meters long and we would have travelled a total distance of 1,073,741,823 meters; that's over a million km,

26.8 times around the world or three times the distance to the moon (when the moon is closest to the earth that is).

When we consider the future we naturally consider it in line with our linear thinking experience. Whilst all of us accept that change is accelerating, that changes are happening faster than ever before, we persist in imagining progress and change as being similar to what we have experienced in the past. We simply find it hard to accept exponential progress; and let's face it, few of us are convinced by the steps example above because well it's a poor metaphor (thirtieth step would be 536,870,912 meters!). We simply have no experience of this exponential growth phenomenon in the real world – or so we thought.

Defining Exponential Technology

In 1965, Gordon Moore, the cofounder of Intel, observed that the number of transistors per square inch on integrated circuits had doubled every year since the integrated circuit was invented in 1958. Half a century later, that observed trend has continued and is quoted as 'Moore's Law'. Moore's extrapolation, that this meant significant increases in the capacity and speed of computers whilst decreasing costs and that this would occur at an ever increasing rate appears to be the primary definition of an exponential technology.

[I'm rusty on Mathematics. I'm a bit confused by this because I've always understood that sequential doubling (2, 4, 8, 16, 32…) was a geometric progression and that an exponential progression was, well a sequence that was exponential, a function or number which had a mathematical exponent [$f(x)^e$, e.g. x^2 – 2, 4, 16, 256, 65536…]. That's probably a bit pedantic; Moore's Law represents geometric growth but I guess we can accept that an exponential rate of change is a type of geometric progression and exponential technology does sound better than geometric technology.]

So, Silicon Valley appears to have led the defining of exponential growth and applied this term to a ubiquitous set of scientific disciplines including artificial Intelligence, robotics, biotechnology, genetics, networks and computing systems, neuro and cognitive science, nanotechnology and additive manufacturing. There remain however mixed interpretations of what constitutes exponential technologies. As Sean A. Hays of Arizona State University posted in 2011[38] "...Ray Kurzweil - and nearly everyone else at Singularity University has taken to referring to an amorphous suite of technologies as "exponential technologies," when, in reality, we only have credible evidence of exponential growth in one field of technologies, ICT..."

He concludes his post with "...Evidence of exponential growth in nanotechnology or biotechnology is virtually non-existent. In the end, the message I would have you take away from all of this is, there is good reason to have hope, but don't believe the hype."

The acknowledged fact that exponential development in ICT will accelerate all other technologies that use ICT (and I believe that's everything) and the mutually supportive cross pollination of parallel advances in all of these technologies is what I understand to be exponential growth. In that sense, technologies so significantly accelerated deserve to be called exponential.

The term 'exponential technology', implying that technologies are developing at an exponential rate may not be universally accepted but when 'thought leaders' create labels they tend to stick. No matter what adjective is used to describe the current technological development frenzy, one thing is clear – it's developing at an astounding rate.

The Basics

I'm writing this book in Australia and like most 'Aussies', I am more familiar with what's happening in the US than the rest of the world; I have studied what emanates from smart 'high tech ecosystems' like Silicon Valley and most of what I've written comes from clever leaders based, or at least associated with, that world. Clearly the US has an abundance of leading edge universities (SU, MIT, Stanford, Harvard, Yale, Caltech, etc.), centres of excellence, private, corporate (Alphabet, Space X, Microsoft, Apple, etc.) and government scientific institutions (NASA, CIA, DARPA, etc.) – simply, many, very bright, well-motivated and resourced individuals in thought communities doing outstanding work.

Writing a book on present technology is a disrupted activity. On a weekly basis, my email inbox is populated with updates and briefs from MIT, Singularity University, the KurzweilAI Newsletter, ScienceDaily and the 'newsrooms' of several globally significant organisations like the European Commission's Human Brain Project, the Korean Institute of Brain Science, Japan's Okinawa Institute of Science and Technology Graduate University, Israel Brain Technologies, Singapore's Brain-Computer Interface Laboratory and China's Academy of Science. Add to that list, old stalwarts like Scientific American and 'breaking news' alerts from a variety of sources and you enter the nightmare of trying to make sense of it all.

This rather topical information overload means that what is written today will often be outdated by the time it reaches readers. Briefing yourself means that one must do, just that. We, if we want to know the latest on anything, have no choice but to hit the internet, select a suitable information platform and read, watch, listen and learn. I'm not surprised that current 'geeks', programmers, hackers and start-up entrepreneurs are increasingly not going to traditional universities or colleges because as they say, curriculums and professors are not up with cutting edge technology and as Elon Musk put it at the 2015 Boao Forum in China, the "data transfer rate is simply too slow". How could it be otherwise? By the time a curriculum and specific subject content has been designed, approved, published and used to competitively secure student enrolments it is outdated before the end of the first semester. Certainly, by the time these students graduate, their mastery on most subjects will be background information on Wikipedia. Traditional education has been and will continue to suffer significant disruption.

An obvious conclusion from all this is that there is so much going on that our linear minds (mine certainly) struggle to digest this fast paced mountain of action driven information. We can however develop sufficient awareness of trends and pursuits that all point to the realisation of 'exponentials' across the entire spectrum of scientific research and the hectic rate at which that is being applied and brought into our experience.

So let's consider the 'keystone' technology that is accelerating all others – computing.

Computers

[Skip this entire backgrounding section if you know about computer development (go to Technological Paradigms)]

From a few beads on a counting stick, to many beads on several parallel sticks (Abacus), from the nearly four century old slide rule to the first analog computers, the pursuit of calculating machines has been a long journey.

The development of computers can be described in terms of generations; each generation representing a quantum leap in technology, paradigm shifts that significantly changed the way computers operated. It's worth noting these generations in outline:

1st Generation [up to 1955]: Electromechanical technology using **switches and relays** replaced by **vacuum tubes**. *Note that there are differences of opinion on what exactly constitutes 1st generation computers; starting from prehistory or the early 1940's, – do you really care?* Some key notes:

- The first electromechanical digital computer, the 'Turing-complete Z3', invented by Konrad Zuse became operational in 1941 and computer science terms such as bits, computer words, clock speed measured in frequencies (the Z3 operated in a range between 5 and ten Hertz), code and data (the Z3 stored these on punched film), floating point numbers and binary systems started to become staples in tech world vocabulary.
- With continuous technological advances (replacing relays with vacuum tubes) and the incentives driven by World War 2, the first

fully electronic, digital and programmable computer 'Colossus' was in service by 1944.

- In 1946, John Mauchly and John Eckert invented the ENIAC I computer which led to the development of the Universal Automatic Computer (UNIVAC); the UNIVAC became the first specifically R&D funded (back in 1946) commercial computer and was delivered to the U.S. Census Bureau in 1951.
- The ENIAC machine used 18,000 vacuum tubes and occupied about 170 square meters of space.

2nd Generation [1955 to 63]: **Transistors**. The bipolar transistor had been invented by 1947 and by 1955 replaced vacuum tubes in these second generation computers which are characterised by:

- Transistors were much faster and energy efficient and allowed computers to become not only faster but significantly cheaper.
- Input was via punch cards and output was via paper printout.
- Programming developed from binary to word based assembly language and evolved to high level programming languages such as FORTRAN and COBOL.
- Development of magnetic core technology enabling computer resident memory.

3rd Generation [1963 to 71]: **Integrated Chips/Circuits**. By 1959 the integrated circuit, a silicon semiconductor, 'integrating' all electrical components, had been invented and patented. Key points are:

- The miniaturisation of transistors enabled 3rd generation computers to utilise ever increasing densities of integration per chip; ranging from small scale integration (up to 100 electrical components, SSI) to large scale integration (up to 100,000 components, LSI).
- Basic computer interfacing changed to keyboard and monitors.
- Operating systems were developed that utilised a central memory manager and enabled the execution of multiple applications (really still sequential but at speeds that appeared simultaneous).

- Computers became more available to an ever increasing number of users.

4ᵗʰ Generation [1971 onwards]: **Microprocessors**. The acceleration point of exponential growth in computing technology had been established. The invention of the microprocessor and particularly the single chip microprocessor (all key computational components now placed on a single integrated circuit, a chip) meant that that earlier vacuum tube ENIAC (with its 18k tubes and a footprint of about 170m²) was outperformed by a quarter palm sized chip. Some details:

- Nearly half a century on, this, 4ᵗʰ generation still defines our present generation of computers.
- The first home computer, personal computer, was released by IBM in 1981; Apple's Macintosh followed in 1984.
- By 1972 Alan Alcorn had already provided a reason for us all to have a computer at home when he released the first popular PC video game 'Pong'.
- Microprocessors were being utilised in an ever growing range of devices.
- The proliferation of ever more powerful and well dispersed computers (personal computers, workstations, multi-user minicomputers, mainframes and supercomputers) facilitated the linking of computers in networks which led to the development of the internet.
- The current era of mobile and multiple application computing, characterised by progressive miniaturisation and a cost reducing and productivity increasing roll-out spiral, had commenced.
- Today's computer is of course hugely more capable then the first commercial desktops. The current fourth generation computers (very large scale integrated circuits, commonly 64 bit microcomputer or Intel 8048/8051 embedded) are the result of exponential development and power our well adapted tools from desktops, laptops, personal digital assistants, tablets to wearable computers.

- Concurrent to the exponential developments in computer hardware there is of course significant complementary development in everything that connects to, is tasked and controlled and interacts with the powerful central processing units of these 4th generation computers. From computer architecture and advanced user-interfacing and software to direct its functioning, everything is in a fast development cycle.

- 4th generation computers continue to be developed in a globally integrated and converging scientific world where 'computer science' now reaches out to and is dependent on cutting edge discoveries in cellular biology, quantum physics, the science of materials, molecular chemistry, nano technology etc.

- The development of weak (or narrow) artificial intelligence and its applications in voice recognition, data analysis, autonomous vehicles etc. is blurring the definition of 4th generation of computers.

5th **Generation** [work in progress]: **Artificial Intelligence**. The quest for ever better computing coupled with the desire to achieve intelligent computers are now propelling the IT industry into the next, the fifth generation of computers. Theoretical and experimental computers are being developed, exploring everything from quantum computing, chemical, DNA, optical, light splitting to spintronic computers.

More on Computers

To get an idea where exponential technology will impact computing and to assist in understanding the basics, it's useful to consider the following key elements. Mind you, a peek at the inner workings of computers is like sneaking a look through a crack in the door only to be hit by a tsunami of detail...

- CPU (central processing unit, the single chip integrated circuit, the microprocessor) – comprises:
  - The central control unit, controls computer components (e.g. add, load, save etc.) decodes programs and generates the signals to activate components accordingly; in complex

CPUs this control unit is in itself a small computer, a micro-sequencer that controls the process.

    ○ The ALU (arithmetic logic unit) executes both arithmetic and logic operations; advanced computers may possess multiple ALUs enabling simultaneous calculations.

    ○ the program counter or register (memory cells) keeps track of where memory

- Memory (computer data storage, semiconductor memory), a collection of 'cells' into which numbers can be placed and which can be 'read'. Cells store binary numbers in groups of 8 bits (one byte) which can represent 256 different numbers, characters etc. (Note that your PC probably has a hard drive capacity, a persistent memory, of say 500 gigabytes (GB) which means it has 500 billion bytes which is 8 x 500 billion bits; that's 4 trillion bits which you could conceptualise as light bulbs being on or off).

Computer memory can be considered from an 'inside out' perspective. The closer the memory is to the CPU the less distance data signals need to travel to get to their control unit defined destination (could be to storage cells or to the ALU for processing) the more efficient, faster the operation and speed of the computer. So from inside out:

    ○ *The Register* is the CPU's own internal memory. Limited in size but containing frequently used data and avoids having to go and get this from outside the CPU.

    ○ Often this register memory is linked to **cache memory** (e.g. L1 and L2, located adjacent to the CPU). This is primarily provided to ensure that the next lot off data is immediately available to the CPU and is in the form of random access memory (RAM – a volatile memory that can be written to and read but will be lost / erased when the computer powers off).

    ○ The next 'closest' memory is the RAM. Again, this can be accessed quickly by the CPU. When a computer is started up it goes through a self-test and setup routine (basic input output system – BIOS) now neatly in the background, before the control of the computer is handed over to the installed operating system. At this stage, the CPU retrieves start-up instructions, persistent data, from the hard drive (usually read only memory – ROM) and brings it into its RAM for quick access and action; ready to respond.

○ The most distant internal computer memory is the *ROM*. This is usually provided in terms of a 'hard drive' and constitutes the majority of the computer's available data storage; usually 100 times larger than the RAM. This persistent memory contains the above mentioned BIOS instructions, the computer's operating systems and any permanently installed programmes or software.

Simple right? The closer the memory is located to the CPU and specifically the micro-processor, the faster it can be accessed and the faster the computer's operation. There have however been some developments that we should note.

Traditionally, data is written into memory by electrical impulses that turn on/off 'micro switches', transistors at the single cell level i.e. each cell stores just the one 'bit', the on/off bit, of information. The invention of 'floating-gate-transistors' has enabled these cells to be made into multi-level and tri-level cells that can store more than one bit per cell. Memory devices that utilises these new transistors are called *Flash memory.*

Flash memory then is a permanent (persistent or non-volatile) electronic data storage device. There are two types of flash memory: NAND and NOR (names reflect the logic gate characteristics of the memory cells). The most relevant here is the NAND type which provides the very useful ability to be written and read in blocks (as opposed to cell by cell) and is electronically erasable and reprogrammable.

The popular USB (universal serial bus) is a flash drive; specifically a solid state (no moving parts) drive that allows us to easily store and transfer data.

Flash drives are gradually replacing computer Serial ATA hard drives (the ROM, drives with spinning platters) with solid state hard drives that are faster and more damage resistant. Other applications include mobile phones, audio devices, robotics, diagnostic instruments – in short any application that accesses data where a degree of robustness and durability is important.

The development of flash memory is also making the delineation between RAM and ROM a little redundant. Remember that traditional RAM was volatile in that it lost its data when power was switched off; well with flash RAM that no longer happens as data is retained so that external to CPU data storage is simply a matter of rapid access rather than storage duration.

Flash memory chips have significantly enhanced embedded computer systems, systems that are function specific and part of a larger electrical or mechanical system e.g. a modern car. The program for an embedded system is stored in its ROM flash memory chip. Often utilising sensors and networks, embedded systems are used in wearable devices (fitness monitors, digital watches), medical diagnostic and monitoring devices and also in large scale applications where processes need to be initiated, controlled or monitored. Typical large scale applications range from flow rate control, traffic management to energy distribution management and can vary in complexity by being networked with multiple microcontrollers and complex peripherals.

- I/O, input and output devices, are peripheral devices that facilitate the interaction with a computer. These devices include anything that inputs data or instructions and receives or acts on the output of a computer or computer systems.

  Initially there were clear distinctions between input (keyboard, mouse, joystick or trackball/pad) and output (cathode ray tube display, printer and speaker) devices. Today, not only are there many more peripherals, they often function as both input and output devices. Some noteworthy I/O devices include:

  - Sensors (activate the computer, instigate computer operation) in response to a particular event or condition)

- Other computer components that input the CPU (e.g. graphics processing units, GPUs, that may in themselves have multiple computers processing and inputting data to the CPU)
- Video and audio devices (cameras, scanners, microphones)
- Interactive display terminals (LCD touch screens)
- Wearable devices (e.g. bio-monitors)
- Data input devices (hard or wireless inputs from network access points e.g. modems/ network interfaces)
- Memory storage and data transfer devices (e.g. hard drives, USBs)
- Multi-media interfaces (e.g. HDMI - high definition multimedia interfaces)
- Any computer, any computer system or network, any web linked device that sends or receives data – in short, any connection

Human to computer interaction is an exponentially growing technology. Continued device miniaturisation, and functionality (a wearable computer watch doubling as a bio-sensor and networked to a PC and connected to the internet is an example) will generate new must 'have devices' at an ever increasing rate. The 'Internet of Things (IoT)', the global interconnection of trillions of sensors, both input and output devices, is already on the imminent to do list and will radically change our human – machine interactions; more on that later.

- Connectors (wires, buses connecting components) both internal between the componentry of computers and those connecting to peripherals have a significant effect on the performance and interoperability. Simple, early computers had, well simple and annoyingly, brand specific, connectors like PS/2 mouse and keyboard connections, a display and printer port and perhaps some audio I/O connections. For networked computers, basic connectors were provided to dial-up access points and perhaps even an RJ45 Ethernet port. Not surprisingly, modern computers now come with very advanced ports enabling much more interoperability and a gradual convergence of compliance protocols. Examples are:

- ○ **Universal Serial Bus (USB).** This is a common and standardised connection protocol (defining the connector, the plug architecture, the associated cabling and communications protocols), connecting standard computer peripherals. USBs are also used on a variety of portable and allied computational and sensor devices such as smartphones, gaming consoles and battery chargers.
- ○ **FireWire.** Whist USBs were primarily developed to provide simple, compatible and low-cost connectors, the FireWire is designed to provide a much faster, high performing connection. FireWire offers high speed, greater bandwidth serial bus connectivity with audio, video and disk drive peripherals.
- ○ **Thunderbolt.** Another peripheral connecting interface, Thunderbolt combines the PCI Express (PCIe, high-speed serial computer expansion bus) with the DisplayPort (DP, digital display interface) and power in a single connector. There is significant effort being applied to developing single cable technology that combines the power carrying capacity of copper wires with the very high data transfer rate that fibre optic technology (tens of Gigabits/sec) can provide. Where connectivity to powered peripherals is desired, non-power optical Thunderbolt connectors are already featured on many computers (it's how most of us connect our Apple computers to display screen/TVs).

Connectors, both internal and peripheral, are becoming key factors in computer and device performance. As microprocessors get ever faster and I/O data is generated at exponentially growing rates, connectors are becoming bottlenecks and are either being developed to achieve comparable transfer rates or to become redundant, to directly link componentry (as in embedded computers).

We can expect considerable development in serial (one by one) or parallel (many at a time) 'bus' (a term applied to any wire or fibre connectors or port) technology. USB, FireWire and similar connectors' performances will remain critical and subject to the chipset, the management of data transfer between processors, memories and peripherals that in turn depend on the bandwidth properties of these connectors.

I started this section by indicating that to appreciate future technological progress in computing we should be aware of what the key components are. We know that the brain of the present generation of computers is the micro-processor, the chip, the integrated circuit. We know that constant improvements have miniaturised this chip and that it can now carry trillions of transistors in any likely configuration of multiple cores and even 3D arrays (remember Watson's componentry) and that these chips control and access at high speeds, vast memories or data banks filled with trillions of information coded bytes. Above all, we know and expect advances in all of the above computational componentry as we approach new paradigms, new generations, of computing.

Technological Paradigms

The above generational view mixes the technologies that delivered them; for example 1st generation computing involved the quite distinct technologies of electromagnetic switches, relays and vacuum tubes. In establishing the exponential nature, the applicability of Moor's law, Ray Kurzweil (Chapter 2 of 'The Singularity is Near') explains these underlying technologies as paradigms.

If we consider any particular technology in the development of computers as a paradigm, (a typical pattern, method, trend, or model) then we have five distinct paradigms. These are electromechanical, relay, vacuum tube, transistor and integrated circuit technologies. Each of these paradigm technologies exhibits a common developmental trend. Ray Kurzweil explains that a paradigm goes through a three stage life cycle as follows. Initially, slow growth, the early stage of growth. The exponential, explosive growth stage and the final levelling off as the paradigm reaches its limit.

As technology progresses through paradigms, these 'S' curves build on each other and can be represented in a continuous curve that cascades these S curves. Clearly, science doesn't generally stand still during the realisation of any one paradigm of technology; rather, as limits become apparent the next paradigm is already being developed. This has been the evidenced trend in the technological evolution of computing and it continues today.

What all of this demonstrates is that a broad technology such as computing is in fact a cumulative set of paradigm technical advances that build on and improve existing technologies. Whilst these paradigms may differ radically in approach or underlying science, their objectives, in this case, computational speeds and capacities are constant.

So what does any of this mean?

Given the performance growth of computing to date we can anticipate its future. The exponential growth curve doesn't stop with the current 4th generation (the microprocessor) of computers or the current 5th paradigm (the integrated chip). Before we look at the future generations and the technological paradigms now being developed consider Ray Kurzweil's 'Law of Accelerating Returns'[39]:

> *"An analysis of the history of technology shows that technological change is exponential, contrary to the common-sense "intuitive linear" view. So we won't experience 100 years of progress in the 21st century — it will be more like 20,000 years of progress (at today's rate). The "returns," such as chip speed and cost-effectiveness, also increase exponentially. There's even exponential growth in the rate of exponential growth. Within a few decades, machine intelligence will surpass human intelligence, leading to The Singularity — technological change so rapid and profound it represents a rupture in the fabric of human history. The implications include the merger of biological and non-biological intelligence, immortal software-based humans, and ultra-high levels of intelligence that expand outward in the universe at the speed of light."*

Note the predictive statement that we will experience something like 20,000 years of progress in the current century. How do we even imagine this; how will this sort of telescoped change 'feel'?

Ray Kurzweil predicts that computers will reach the level of a mouse brain in the current decade, the human brain in the early 2020s and surpass the capacity of all human brains combined in the 2050s. Now, you might not like or accept these predictions but if you accept the rate of progress to date and the nature of exponential growth this has established and you also understand what is being developed today, then you may need to, at the very least, accept that this is where we are heading. You would need to acknowledge that these advances are achievable and even if the timelines turn out to be more fluid (could even be plus or minus a decade or two) there is little ground to not acknowledge final outcomes.

In considering the progress of technologies it's also important to recognise how consequential changes are manifested and experienced I our daily lives. Why is it that we hear about breakthrough discoveries and then nothing seems to happen? Why does change appear to be so rapid sometimes and what is this 'disruption' all about?

The Six D's of Exponential Technology

In '*Bold*', Peter Diamandis and Steve Kotler introduce and explain the following 'six D's' which describe a progression of phases that apply to the life cycle of pure and applied exponential technologies:

- Digitalization [Digitisation]
- Deception [Disappointment, Disillusionment]
- Disruption
- Demonetization
- Dematerialization
- Democratization

Let's look at each step.

"Digitizing or digitization is the representation of an object, image, sound, document or a signal (usually an analog signal) by a discrete set of its points or samples. The result is called digital representation or, more specifically, a digital image, for the object, and digital form, for the signal. Strictly speaking, digitizing means simply capturing an analog signal in digital form."

This first phase simply represents the fact that when anything is digitised it becomes subject to the analytical, manipulative application of exponential ITC power. A much quoted example of this is the demise of Kodak (Eastman Kodak Company).

How did the digitisation of the 'photograph' by an employee (Steven Sasson, the inventor of the first digital camera in 1975, working in Kodak's Apparatus Division) lead to the collapse of the world's leading film and camera company; a company that in 1996, employed 140,000 people and had a market capitalisation of 28 billion dollars?

Clearly, the digital camera transformed the capturing of an image, the photograph, into a digital format. This then allowed the image to be manipulated, transmitted, shared, stored etc. using ITC. As an aside, Instagram exploited this potential brilliantly combining the properties of an instant camera with the transmitting characteristics of a telegram (hence 'Instagram'). Instagram launched as a free mobile app in 2010 and in April 2012, with only 13 employees, was acquired by Facebook for around 1 billion dollars.

Photography had been *digitised* in 1975.

The time from the invention of the first crude digital camera (1975) to the commercial integration of cameras in mobile phones (first mobile phone camera was sold in Japan in 2000) – can be viewed as the *deception* phase in that there would have been an expectation, anticipatory hype, that the world would soon enjoy the benefits of digital photography but they had to wait some 25 years for its widespread realisation.

Kodak was terminally **disrupted** by the digitisation of photography. This disruption was a long and torturous process. It highlighted that size, financial clout and even leading edge specialist technology and product excellence was not sufficient for Kodak to survive the disruption once exponential growth in the new digital imaging technology took hold. The disruption that caused Kodak to fail was an **opportunistic disruption** for Instagram and many more commercial entities that gained from related technologies (image manipulation, medical imaging, CGI, visual media, mobile phones etc.).

Peter Diamandis explains disruption in terms of the gap between our linear perceptions, our natural step by step mental appreciation of events and the fast paced 'exponential' growth of technology.

This gap, between these competing and diverging events, is what causes the disruption. The speed and potential effect of such disruption can manifest as a destructive disruption (a new 'Kodak Moment') whilst simultaneously presenting as an opportunistic disruption (the 'eureka' moment for the inventor in a garage or a start-up entrepreneur).

Exponential technologies also have the effect of saving on materials and even negating the requirement for materials – that's **dematerialisation**. Again, considering photography as our 'lab rat', no more film developing chemicals or photographic paper (disregarding remnants in the film industry or artistic uses), no more industry related logistic supply lines.

Perhaps not the most breathtaking dematerialisation example but it's quite significant when you consider paper saved in electronic correspondence. Saving on actual materials also includes no longer producing products like (look around a new home) clocks, typewriters, radios, atlases, sets of encyclopaedias and associated furniture.

Another aspect of this phase is the dematerialisation of services that we once had to 'go to' that we now have at home. From the cinema to video hire, record store to amusement parks and video game parlours, the printing shop, the post office, the information counter for products or services or simply running around to find out this or that, finding the best deal (all variously disrupted industries or activities now) most of these are now available at minimal cost with much more choice, customisation and scope, in our homes. Exponential technology does not just dematerialise materials; it disrupts and often dematerialises whole sections of commercial and industrial enterprises and frees up resources and a lot of our time.

Returning to the digitisation of photography, few would disagree that photography has become almost costless. It costs little in energy to take, retake, send, store, alter, enhance and share digital images. That is demonetisation. No more money spent on films, no expenditure of time waiting for processing, cost savings in immediate quality feedback (look and snap some more), albums and no deterioration of images.

Demonetisation takes money out of consideration. Whether this is brought about by a new technology replacing what used to cost money (the free digital camera in your phone) or the numerous new services via the internet (the internet itself, search engines, emails, Wikipedia, news, weather, maps etc.) there is no doubt that they have all eliminated direct costs for 'things' we used to buy. Whilst we are instead trading information about ourselves, our habits, behaviour and interests and populating huge data as consumers, we are nevertheless benefiting from the rapid growth of demonetising technologies.

The last phase in the exponential technologies cycle is ***democratisation***. Basically, as a product or service demonetises, gets cheaper and cheaper, it becomes universally affordable i.e. democratised. Assisted by exponential growth in sales, ICT assisted scaling; costs reduce to near negligible levels. Self-explanatory really and computers, tablets, mobiles are good examples of democratised technologies.

There is nevertheless a price to pay. Someone is paying for many of the free products and services that exponential technologies continue to provide or enable. As I've mentioned before, we pay in providing ever more detailed information about us. That information is generating significant incomes and valuable consumer data to the providers of free services.

To what extent new innovations of the future follow the above 6D cycle is perhaps academic; what is however absolutely clear is that these phases have been evidenced to date and are spreading across an increasing range of industries and enterprises. It's not a big stretch to conclude that those enterprises that appear to be immune to disruption today, those not dematerialising and demonetising will for that very reason become ripe targets for disruption, both opportunistic and destructive.

Disruption is Disruptive

The term 'disruption' is being used more and more by business analysts, leaders, commentators and observers. Once just a term of annoyance, perhaps something interrupting the flow, a process or activity, today it means massive impact, anything from a change of emphasis or direction to the potential cessation of entire sectors of enterprise and industry. A disruption can be slow acting or if as a result of exponential technology arise like a snake bite, seemingly from nowhere and potentially fatal.

When most of us consider disruption we focus on the above Kodak example. Perhaps the joke that it had its 'Kodak Moment' or the fact that such a large corporation could in fact be put out of business, disrupted, makes it such a popular example. But there is disruption everywhere and we, just like nature, don't appear to mourn extinctions.

Here are some disruption examples [and what changed] that most of us don't care much about anymore:

- The word processor and the personal computer [massive reduction in office administrative staff, typing as a profession, decrease in secretarial work, registration and document handling]
- Internet and email [communication industry, postal services, newspapers, magazines, libraries, encyclopaedias, atlases, maps, education (online courses) etc.]
- Mobile and smart phones [telecommunications industry; in fact smart phones disrupted major previous phone tech companies in that it replaced cell phones, palm devices, cameras, portable music

players, sound recorders, directories and many other services that are now provided through specialised applications etc.]

- Recent computers [modern lap tops and even desktops have continually disrupted their earlier iterations; un-noticed by us perhaps but not the industries forced to continuously re-tool and keep up]
- Software [from operating systems to gaming, programming efficiency and interface quality – all continually improved to take advantage of hardware capability increases; industry continually updates or becomes extinct]
- Social networks [disrupted all forms of networking, communicating and socialising, covertly disrupting privacy, the collection of data (establishing habitual behavioural data bases), continued the disruption of information services, disrupting confidentiality and information restrictive paradigms]
- The cloud [disrupting major tech hardware companies such as IBM, industries providing hardware and data hosting and management services, replacing in-house data storage providers, disrupting privacy and data security industry]

These are just a few; ask any 'older' person what they no longer use in their house and you'll find many more apparently insignificant items and tools that are no longer required. Pens and paper are among those for many. We more often just hear about the newsworthy disruptions such as:

Industry/Technology	Disrupting/Disrupted
Google (Alphabet)	Libraries, research, Information providers
iTunes	Record stores
Uber	Taxi Industry
Amazon.com	Book stores (e- books - book production and distribution)
Skype and Facetime	Telecommunications (particularly long distance)
airbnb	Hotel, motel industry
craigslist	classifieds
ebay and alibaba	Traditional retail stores
Crowdsourcing e.g. Freelancer, topcoder	Agencies, human resource brokers, job search providers, specialist service providers
Crowdfunding	Investment brokers, industry investors, traditional financing industry
Bitcoin	Banking and financial industry

If you look at those companies and industries that have been or are being disrupted what might we conclude? Is their failure due to poor performance or bad product? No, not usually. They fail because they didn't or couldn't adapt to new technologies.

What is indisputable is that exponential technology will disrupt and disrupt mercilessly. It will continue to be an entrepreneur's heaven, providing massive opportunities with overnight success potential. The flipside for linear thinking, people, inflexible and outdated enterprises, corporations and even government instrumentalities is well - quite grim. It really is prudent to accept what's been happening, to learn and adapt. Change is not new; it's the rate of change that is so massively disruptive.

More than 250 years ago, the developed world commenced an era of massive transformation. This transformation, the Industrial Revolution, advanced technologies that affected everything from manufacturing, materials, resources, energy, transportation, communication, labour reforms and even politics (socialism, capitalism and romanticism). It was also an era of innovation and knowledge and it has continued to transform all aspects of human endeavour to the present day. Since the emergence of computers, six decades ago, technology, and in particular, information and communications technology (ICT) has accelerated this transformation and brought us to the 'Information Age'.

Information enabled technology, once digitised, demonetized and democratised, is fuelling exponential growth in innovation which is a defining feature of this significant age.

The information age is well and truly here but it is still at a very early stage. The relentless growth and global spread of exponential technologies is about to severely test us all. The extent to which governments, corporations and individuals will be able to cope with the inevitable changes that are about to manifest will depend on leadership; leadership into a highly connected, interdependent but very vulnerable and unpredictable world. Only one thing is certain – 'disruptive change' is the overarching paradigm of the information age.

Alan Murray reporting on discussions at 'Fortune's Brainstorm Tech 2015' event held in July 2015 in Aspen, Colorado wrote[40]:

"Here's my takeaway from day two of Brainstorm Tech: Technology is not the answer". He explains that what is required is cultural change and leaders with the skills to adopt, develop and utilise rapidly evolving technologies. Certainly, Alan Murray is correct in what he and his colleagues have concluded. The commercial world will indeed require exceptional leadership but – and this is a huge but – so does every governing body around the world.

So, we've looked at the meaning of exponential technology, its stages and disruptive impacts and we've looked at the main enabling technology, computing. We've really just covered the machinery, the hardware of computing. What we do with this hardware, how we task it through instruction, the software, is equally important.

Convergence of Science and Technologies

The exponential realisation of computational capacities, the availability of faster and cheaper computing power is having a convergent effect on all other technologies. The exponential, Moore's law abiding development trends are manifesting in the development of a range of technologies which in turn 'infect' other technologies in mutually beneficial feedback loops.

The four most significant technologies that I will detail in the following chapters are:

- The Next Paradigm, the 5th Generation Computers, Networks and Sensors
- Artificial Intelligence and Robotics
- Nanotechnology and Digital Manufacturing
- Genetics - Synthetic and Digital Biology, Medicine and Neuroscience

CHAPTER 7

Computing, Networks and Sensors

The Next Computing Paradigm

The current, 5th paradigm of computing (the Integrated circuit) has been around since the mid-1970s and will most likely reach the top of its S-curve in this decade. In other words, it will soon reach the limit in its exponential price/performance growth. As previously pointed out, the exponential nature of this technology implies that the next, the 6th paradigm, should by now be well in development if Moore's law is to hold true into the near future.

Not surprisingly, the next innovation leap in computing has been worked on since the 1980s and the primary focus remains on 'quantum computing'. In the previous pages I mentioned the convergence of science and technologies and the research and development effort in computing evidences this clearly.

The developmental effort appears to be leading towards three dimensional (3D) molecular computations. Replacing the fundamentally two dimensional computer chip with a three dimensional architecture makes sense - simply more volume but it is the molecular characteristic that is the game changer. This is why the targeted 3D molecular computation chip might be achieved through quantum, biological or optical computation technologies.

These technologies are 'high tech' and to comprehend these beyond the very basic level, a sound understanding of science and in particular physics, chemistry, biology – well science actually – is necessary.

[Rant: I recall my own 'high school' experience where out of 120 students only three of us studied 'Level 1 Physics'. At that time, most thought that nuclear physics (even Newtonian physics) was absolutely pointless and would not only be too difficult to learn but obviously unnecessary. So whilst very few of us even recall the Bohr Atom, Schrödinger's Equation, the Heisenberg Uncertainty Principle or the three page derivation of $E=mc^2$, the majority of people have no understanding of sub molecular structures, atomic and subatomic particles and the properties of quantum physics. They will understand molecular computing – how? I will come back to this point when we look at education in Part 3.]

The above three likely computational paradigms will involve today's developing technologies that include the apparently different but convergent technologies of nanotechnology, biotechnology and subatomic particle physics. This convergence is logical as our very categorisation of sciences and technologies break down when we shrink our focus to the molecular, atomic and sub-atomic level. Everything is made up of these elementary particles, it is their combination, intelligent cooperation and resulting function that lead us to label them and perceive them as being 'different'.

An important take away here is that the next computational paradigm will draw heavily on a variety of technologies, convergent and mutually accelerating. These will be any combination of Genetics, Nanotechnology, Robotics and Artificial Intelligence.

Quantum Computing

This appears to be the technology most likely to define the next generation of computers. The theory is well understood; its practical realisation is being developed. So what is quantum computing?

Traditional computing, all previous paradigms, have basically made use of an on/off switch to store information. The most basic form of this is called a 'bit' which stands for 'binary digit'. The binary system is a mathematical base 2 methodology that allows any number to be represented by a sequence of zeros and ones; on and off states of a switch. In present computers theses switches are transistors embedded in silicon and connected to create logic gates to process information.

When we describe computational memory we cite numbers of 'bytes', one byte commonly consists of 8 bits. So if you've got a 1 gigabyte memory stick you effectively have 8 gigabits or 8 x 1,000,000, 8 million bits or 8 million 'switches'. Computations calculations, computations, manipulate these bits to carry out mathematical operations. In the example of IBM's 'Watson' [see Chapter 2] winning at Jeopardy; 'he' could process 4 trillion bits of information per second (500 gigabytes).

The aim of quantum computing is to replace the above 'bit' with the 'qubit'. A qubit employs quantum mechanics which means that the bit can not only be on or off, represented as 0 or 1, but that it can also be both at the same time. This qubit is in a state of superposition and can therefore take computational 'shortcuts' and radically outperform traditional bit based computers. Not a simple concept to understand I know as it reflects the wave-particle duality of matter [research quantum physics to understand this].

Quantum computers aim to utilise quantum mechanical properties of fundamental particles in an array of qubits to represent and organise information that can be processed. There are however major difficulties in managing the uncertainty issues with the 'quantum states of qubits' and their grouping into useful sized arrays. Whilst there are many that have solved the underlying architecture of a quantum computer, technical challenges are being addressed around the world in what has become an international race to produce a commercial quantum computer.

Google has recently claimed that it, in collaboration with NASA, has developed the D-Wave 2X quantum computer which reportedly is 100 million times as fast as currently available computers. There are numerous governments, research labs and even start-ups working on competing, complementary and alternate approaches to create quantum computers but for now, Google appears to have the edge. Whilst significant challenges persist, technology and the bridging (new technologies bridge capability gaps that develop other technologies which in turn bridge gaps and so on) through the development of supporting technologies is approaching the practical realisation of massively more powerful computing.

Quantum computing offers the technology that would enable computers to calculate, process information, in seconds that would take the current generation of computers thousands of years to crunch.

Molecular Computing

The computer of the future will most likely have architecture at the molecular level. This could be a biological brain mimicking arrangement, a nature based DNA arrangement or a Nano technological arrangement made up of single layers of atoms and molecules. At this stage there are various 'nuclear behaviours' and sub-atomic particle physics properties that are being investigated that might manipulate the evolved bit, the qubit, and produce super-fast computing devices.

One possibility is optical computing (also termed photon computing) utilising the fundamental particles of light, photons, to replace the electron driven transistors and semiconductors that are used today. It is likely that future computers will mix various technologies to provide specialised computational abilities. Technologies such as spintronics (the spin of electrons), photonics (the splitting of light, laser photon streaming) and the self-replicating and massive code bearing properties of proteins (DNA, RNA) are likely to be used in conjunction with nanomaterials such as nanotubes (carbon molecules arranged in a seamless tube) and graphene (single virtual one-dimensional layer of carbon) to advance computing in the next generation. These future computers are likely to mix organic biological functions and inorganic materials (silicon, carbon) that will use less power, generate less heat and operate at the speed of light. This is all work in progress.

Whilst we've been able to look back through the now very logical paradigm shifts or generational developments of computer technology the future of computing promises to be very complex with equally huge quantum leaps in both processing speeds and memory capacities.

This begs the question of just how fast do we want or indeed need computing to be. It will also become important to justify the level to which such powerful computing needs to be distributed throughout a population. Whilst it easy to appreciate the benefit of super quantum computing in cypher security, massive data storage and applications and central processing technologies like a future global search engine or cloud based medical know-it-all (like 'Watson') it's not so clear that we will need that sort of performance on our laptops. Of course I'm not anticipating a lap top as such, I'm imagining some retinal implant, a wearable computer – well a ubiquitous computational device yet to be determined but it will be highly mobile and fundamentally unobtrusive – I think.

Ubiquitous Computing

'Everywhere' or 'all over the place' computing is more eloquently termed ubiquitous computing. This ubiquitous computing concept combines all forms of computational devices with how and where they can be used.

The forms of computing range from huge mainframes to desk and lap tops, to tablets, wearables and extend to object imbedded or attached context and environmentally aware devices. In this context, the definition of 'computer' broadens to include any device that senses, measures, collects data, interprets, activates, stores or transmits information, responds to instructions (real time or predetermined) at various levels of complexity and sophistication.

An example is the domestic fire/smoke alarm. For most of us, it's simply fixed to a wall and provided we've kept the battery in order it should on detecting smoke emit an audio alarm. It could and should however do much more. It could for example do more to wake or alert us if it has detected that there is in fact a life form in the space that it's protecting. It could flick lights, alert neighbours, activate emergency response calls, commence fire suppression, release oxygen masks (as in aircraft), manage air movement (ventilation, smoke exhaust) and even physically provide escape mechanisms. You can imagine all sorts of things that an alarm might do, and such devices are already being progressed.

We are all witnessing the migration of computing technologies into ever more mobile, smaller and unobtrusive devices. We can now do more on a wearable computer watch or a small handheld phone than what used to require a large desktop computer. The form of computing is clearly trending towards small, highly portable, affordable networked devices; not necessarily as replacement but primarily to augment and achieve the 'everywhere available' objective.

The technology that enables this pervasive ability includes the internet, the cloud (remote accessible data storage), various operating systems, programmes such as pattern recognition and natural language (multilingual), mobile technologies and protocols, positioning/location systems, sensors and the internet of things and computer componentry (microprocessors and various input/output systems). We will look at some of these in the pages to follow.

The final factor enabling ubiquitous computing is user interface. No longer will users be tied to a fixed screen, regardless of size. Users will be able to 'see', hear and generally perceive and interact with computers in a variety of ways. It is likely that these will include any surface and even 3D holographic displays and input systems; these can be at microscales (lenses, glasses, retinal projections) or large cinematic display screens.

When you consider the progress in exponential technologies it is clear that the three fundamentals driving the ubiquity of computing are all being developed aggressively and are mutually enabling and building on each other.

Networks, Sensors and the Internet of Things

Ubiquitous computing, even at the level at which it exists today, would not be possible without the ICT network. Today's hyper-connectivity is a result of rapidly developing technology in hardware, the networked machines, the software to drive them and its connectors. Let's have a quick look at some key elements of IT networks:

Internet

The Internet is a global network of interconnected computers and computer networks. It is the infrastructure of hard and soft assets linking transmitters and receivers of electronic data.

To enable networked computers to communicate in an end to end process, a suite of standardised conventions was adopted. These conventions, the Transmission Control Protocol (TCP) and the Internet Protocol (IP) define how date communication is to be transmitted, organised in sequentially structured 'packets' and how it is to be routed to specific recipient addresses.

A feature of the internet is this unique IP address system. It is the IP address, expressed as a set of four numbers (e.g. MIT has the IP address: 18.72.0.3), which actually enables the internet by allowing computers to locate and identify each other. Currently, the internet uses Internet Protocol Version 4 (IPv4) which has an address limit of 4.2 billion unique addresses.

All available IPv4 addresses have now been allocated. To overcome the future need for more addresses, (growing number of internet users and devices that need identification) IPv6 is currently being developed. This much larger system will provide a maximum of 3.4×10^{38} or 3,400 trillion septillion unique addresses. Whilst this seems a bit like overkill, it will adequately provide for unique identifiers of all 'things' to be addressed in the emerging 'Internet of Things (IoT)'.

WWW

The World Wide Web (www) is an internet application. It is a collection of web sites, often consisting of multiple web pages that are interlinked as hypertext documents - web pages. Web pages are commonly written and formatted using Hypertext Mark-up Language (HTML) and often include text, visual and audio media, navigation bars, sidebar menus and embedded hyperlinks to web pages on the same site or to external web sites.

Browsers

Browsers are application software that allows users to access and utilise content on the World Wide Web, private networks or data/file systems. Browsers use Hypertext Transfer Protocol (HTTP) to communicate and access information requests. Accessible information resources are identified by browsers via a Uniform Resource Locator or Identifier (URL or URI). Common browsers are Google, Safari, and Internet Explorer but there are literally hundreds including the Russian Yandex and China's Baidu which are emerging in global relevance.

Search Engines

Search Engines are software systems with efficient algorithms to locate and present specific information on the World Wide Web. Search engines have solved the initially frustrating problem of trying to access information or even to know what is available. User feedback loops, usually passive and ambient, are increasingly used to establish individual search habits and to mediate what is offered as a search result.

Whilst I'm on searching the internet, you might be surprised to know just how much of the internet and specifically the web you actually have access to when you use the world's largest and most powerful search engine – Google. Well, you can actually only access 0.03 percent of what actually available. In other words, if all the internet information makes up 10,000 pages you have access to 3 pages. The majority of content is in the Dark

Web, the domain of the hidden, which is more than 300 times larger than the visible web.

Most search engines are 'free' and you might correctly suppose that there is a price. That price is loss of privacy and the massively purposeful pursuit of data on everything. Search engines mine data, read and store all information and whilst this is perhaps necessary (how can you be served what you want without the searcher knowing where, and exactly what, the information is?) you should be aware that engines such as Google and Baidu do collect data on all of its users.

According to Marc Goodman (in his book 'Future Crimes'), "Google reportedly has at least fifty-seven separate personalisation signals it tracks and considers before answering your questions, potentially to include the type of computer you are on, the browser you are using, the time of day, the resolution of your computer monitor, messages received in Gmail, videos watched on YouTube, and your physical location." Nothing particularly disturbing there unless you're wondering what happens to your profile, how and who is managing this and what happens to all that data on you. Rest assured that the data will never be erased and be aware that you are trading privacy for the enjoyment of instant and efficient access to global information.

Telecommunications

Telephone Networks and in particular our mobile networks, remain hugely important and are becoming a 'must have' communications facility. The mobile data protocols adopted, the generations (G) of wireless internet technologies that progressed from the 1G analog, through 2G early digital to the present 3G and 4G are delivering astounding mobile communication capabilities. These networks provide the infrastructure and protocols for wireless internet access to smartphones, computers, IP based telephony, to cloud computing – basically it transfers traditional computer capabilities to the networked smartphone.

We will see continued development of wireless broadband mobile technologies including adding to the already several million useful, and sometimes trivial, applications available. Work on 5G technology is well underway and promises to deliver a network capable of increased data rates, simultaneous connections that can handle large numbers of sensors, better coverage and improved all around efficiencies. The anticipated roll-out of 5G will be highly dependent on the implementation of the before mentioned IPv6 address system.

The connectivity provided by a global network of undersea cables, satellites and reticulated wire or wireless connections is the infrastructure that has revolutionised communication. Our utilisation of this infrastructure is however governed by the Internet Service Providers (ISP) we choose or are forced to use. The internet is the common denominator of virtually all modern communication. From telephone to email and data transfer, everything is controlled by networked computers, everything is rendered as digital or analog information, it is all highly collectable, storable, subject to analysis and therefore useful.

What you connect to, what you access is at the discretion of search engine providers. Whilst some search engines simply collect user data on you, others may impose restrictions and filter what information you can or can't access (i.e. the much talked about Great Firewall of China). It is worthwhile to note that there are only four nations that have their own core search engine technology: the US, China, South Korea and Russia.

It is becoming increasingly apparent that our technological advances in networked communications have created very powerful information and influencing tools. Networks are influencing politics, business, industry, trade and global alliances. The implications of IT hyper connectivity, the wide ranging effects of enabled immediacy of event awareness and factual knowledge, will define our global future.

Sensors

The obvious definition of a sensor is any device that detects something and generates a corresponding response or output. We've all been exposed to basic sensors that detect motion (security lights, wearable activity sensors, thermometers, touch activated switching or dimming of lights, touch pads and even the keyboard of our computers is touch sensitive). Fundamentally a sensor reacts to change and that change can take many forms. Wikipedia lists a fairly complete range of sensors[41]:

- Acoustic, sound, vibration (geophone, hydrophone, lace sensor a guitar pickup, microphone)
- Automotive, transportation (many sensors from visual blind spot monitors, vehicle performance, speed sensors, critical fluid monitoring etc.)
- Chemical (gas/air sensors, smoke/pollutant sensors even the policing tool the breathalyser)
- Electric current, electric potential, magnetic, radio
- Flow, fluid velocity (sensors that measure the flow and flow characteristics of gas, liquids and material)
- Ionizing radiation, subatomic particles (e.g. Geiger counter, neutron detector)
- Navigation instruments (air, land or sea travel related instrumentation)
- Position, angle, displacement, distance, speed, acceleration
- Optical, light, imaging, photon
- Pressure
- Force, density, level
- Thermal, heat, temperature
- Proximity, presence (alarms, radar, motion, occupancy, proximity, infrared and switches)

I detail the above because it is important to appreciate just how much we are already using sensors in everyday life. With the miniaturisation and inclusions of computing capability, microprocessors, in sensors they are increasingly becoming more useful. They are increasing their capacity to 'act' on what they are sensing; that is they are being enabled to output electrical or optical responses to detected changes in exponentially more complex and value added ways.

Sensors are becoming more prevalent in industry, infrastructure grids, healthcare facilities, logistics and remote services in that they, with the appropriate connectivity provide real time condition monitoring to everything form an implanted medical device to the edge condition of a 30 meter long wind generation blade or remote oil pipeline. Condition monitoring of any mission critical structure or machinery can be inexpensively provide by sensors that operate around the clock. What makes sensors particularly useful is when they are networked and connected to be managed in a ubiquitous manner.

The Internet of Things (IoT)

I introduced IPv6 above (3,400 trillion septillion unique addresses). It will be this availability of virtually unlimited but unique identifiers that will connect and network sensors and machines with all the data handling possibilities that characterise the internet.

The IoT will bring to everyday things, machines similar capabilities we now associate with smart devices like our latest mobile phones. This IoT will utilise cloud computing and join the growing network of data gathering and reporting sensors and it will facilitate software driven machines, sensors, communicating with each other. As IP addressable devices these connected 'things' will join the ubiquitous technologies in mobility and connectedness.

The most significant development that the IoT enabled sensors bring will be the ability to gather information, data, and to use it. This leveraging of data will utilise cloud based applications, IPv6 enabled identity management and massive processing to interpret and transmit, act on, the sensed, detected and reported information.

We can all imagine what living in an 'Internet of Things' world will be like; anything worth measuring, observing, collecting and analysing will be processed. What will your smart home and its smart utilities monitor and report? Will appliance share power, turn themselves off in a response to environmental changes, will your air conditioning create what it knows to be your preferred and mood scented micro climate? How about your smart office? Will your computer interface report a lack of attention or poor performance or your chair record your occupancy? What then will your smart city be like? Will you need to buy tickets for anything or will transportation monitor your movements and charge your financial institution accordingly?

The IoT is fuelling huge interest in business. Its promised massive data collection abilities do appear to present opportunities in matching products and services to individual's evidenced behaviour and preferences; all enabled by sensors of the Internet of Things.

CHAPTER 8

Artificial Intelligence and Robotics

The Horsemen of Technology

The three most impactful sciences and applied technologies, those that are likely to have near future implications for all of us are undoubtedly:

- Artificial Intelligence and Robotics
- Genetics (Synthetic and Digital Biology, Medicine and Neuroscience)
- Nanotechnology (and Digital Manufacturing)

Of course these are driven by computing and all allied technologies as well as the general convergence of mutually supportive scientific disciplines. Nevertheless these three could be described as the apocalyptic horsemen of technology. I use apocalyptic in the sense that they will enable a super-accelerated transcendence of human limitations. And as with all powerful technologies this will be both enormously beneficial and unfortunately, just as significantly perilous.

In earlier chapters I introduced Ray Kurzweil and referenced his 'The Singularity is Near' as a publication that I had studied. I confess that I mostly accept its contents. Kurzweil defines 'the Singularity' as a "…future period during which the pace of technological change will be so rapid, its impact so deep, that human life will be irreversibly transformed."

He explains this singularity as "...neither utopian nor dystopian" and that "... this epoch will transform the concepts that we rely on to give meaning to our lives, from our business models to the cycle of human life, including death itself."

Opinions about the Kurzweilian singularity vary widely and are today just as passionately analysed and discussed as back in 2005 (first publishing of "The Singularity is Near'). You may intuitively simply not accept anything like a singularity event in our human future; indeed there are many that believe that there is a greater likelihood that we will by our own hands, become extinct as a species before such technologies materialise. Perhaps, like most of us, you've never really looked or deeply thought about where all our innovations and technologies are leading us.

As you read the coming chapters I'd like you to recall the earlier mentions of purpose, higher purpose, values and how their relationships are interdependent on the perceived reality (the objective and artificial, human constructed realities) that define life for us individually and collectively. Just one of these human constructs, say that of 'employment', serves as an example of how technology might actually *transform* one of *the concepts that we rely on to give meaning to our lives.*

Without belabouring the point I think we all recognise the 'what are you going to be when you grow up', 'what do you want to be', and 'what do you do (for a living)' as social imperatives and most of us share the history that has defined who we are, what we have and how successful we've been, in terms of our employment, in terms of what we did.

The Swiss held a referendum[42] (June 2016) on a proposal to introduce a basic income for all citizens. The referendum proposed an unconditional basic income, a non-means tested income for everyone. The referendum sought to address poverty and unemployment. It would have provided every adult and child with sufficient income to lead a dignified life, employed or not (proposed as a monthly stipend of 2,500 francs per adult and 625 francs per child).

The referendum did not succeed but it is an exploratory step towards a major paradigm shift; a radical change to what has in the past projected a social and industrial 'value' onto humans. Today, a number of nations are exploring the introduction of a non-employment based 'living' salary.

The Swiss example is just one issue that points to changes that are approaching. Certainly we are hearing more often about the concept of living salaries being considered by a number of nations. Given the massive global inequalities in wealth, rising unemployment and the technologically underwritten trend for much greater un-employment we should accept that life will see many more paradigm shifts and perhaps help us to at least agree the broad future direction that Kurzweil predicts. I'll come back to the future of work and employment in Part 3.

What then have our apocalyptic horsemen of technologies to do with unemployment and how will their technologies shape and define our future? To answer that question we need to understand a little about these technologies and their current and expected impacts.

Robotics

In Chapter 2, I introduced artificial intelligence (AI), what it is, the 'Turing' test, evidence of apparently intelligent machines (IBM's Big Blue and Watson and Google's AlphaGo) and some initial human brain to computer comparisons. Given that the human brain is the most intelligence enabling organ that we know, it wasn't much of a stretch to conclude that it is indeed worth copying in our attempts to make machines intelligent.

We do already have what many consider intelligent machines (other than computers) – robots! Most commonly, for now, robots are electro-mechanical machines, virtual artificial agents or devices (they could be biological or chemical hybrids) that are controlled directly (including brain-computer interfaces), by electronic circuits, through programming and can operate semi or fully autonomously.

Mechanical robots range from anthropomorphic (humanoids) to industrial, service, special purpose robots to self-organising swarming robots. Robots to varying degrees emulate life; they can be mobile or static, operate in air, sea and land environments at levels surpassing human limitations, they can operate in a full range of modes (degrees of freedom) and basically take on any shape; they are being built to mimic known life forms from microscopic to large scales.

[*Note that there are also non mechanical robots termed 'bots' that may be software in form and are defined by their task automation such as web crawlers, internet bots, chatterbots (Siri using natural speaking software) and even computer game opponents.*]

Whilst we are all familiar with robots in humanoid form, today's robots are also mimicking a wide range of animals. This range of bio-inspired robots (biomimetic) includes:

- Big dog – a self-adjusting and athletic four legged 'dog like' robot that travels across a range of difficult terrains
- A cheetah that can anticipate the need to jump over obstacles and successfully do just that
- A hopping robot kangaroo – how is that for robotic balance
- Robotic fish (well – they swim)
- Robotic spiders (hexapods) that can not only walk like a spider but can transform into a ball and roll
- Robotic birds and insects that fly and spy and can be life realistic in size or as small as a bee (includes swarming robots that self-organise and cooperate to complete tasks)

Just imagine a swarm of bees; let's arm them with lethal stings, (say a genetically engineered Ebola/influenza combined virus or simply a fast acting deadly poison) and show them a picture of a particular human target. Given that each bee has artificial intelligence and can identify, 'lock on' and determine effective attack strategies (all possible already) and that this swarm of say 20 bees will self-organise and coordinate their attack - what are our target's chances of survival? Sadly not science fiction!

What about taking existing bird robots, say one that looks, flies and acts like a pigeon and let's task it to visit the nearby high security prison. Could we get it to deliver heroin, explosive material or even miniature weapon components? Of course we could and with technology that already exists.

[*Rant: I've had to listen to someone recently assuring an audience that robots will never be able to harm humans. He explained that the well-known science fiction writer, Isaac Asimov, had, back in 1942, already dealt with this issue. He then referred to Asimov's three laws:*

1st - A robot may not injure a human being or, through inaction, allow a human being to come to harm.

2nd - A robot must obey the orders given to it by human beings, except where such orders would conflict with the First Law.

3rd - A robot must protect its own existence as long as such protection does not conflict with the First or Second Law.

He didn't mention the final, fourth law added later: (4th - A robot may not harm humanity, or, by inaction, allow humanity to come to harm).

So I guess we can all relax because as we know when we have laws – everybody obeys.]

I mention the above only to make the point that with every perceived good or apparently frivolous application of these emerging technologies there is a corresponding risk of misuse. The laws of Asimov above can actually be implemented for narrow, weak AI robotics; the problem however becomes gradually more difficult as AI becomes more powerful and once AI surpasses the capacity of the human brain it may prove impossible.

So, how do we define a robot?

The International Federation of Robotics (IFR) classifies robots as either Industrial or Service Robots according to their intended application. This reflects the international standards definitions (ISO 8373):

- An *Industrial Robot* is "An automatically controlled, reprogrammable, multipurpose manipulator programmable in three or more axes, which may be either fixed in place or mobile for use in industrial automation applications"
- A *Service Robot* is a robot that performs useful tasks for humans or equipment excluding industrial automation application. It is further defined[43] as being:
    - A *personal service robot* or a service robot for personal use is a service robot used for a non-commercial task, usually by lay persons. Examples are domestic servant robot, automated wheelchair, personal mobility assist robot, and pet exercising robot.
    - A *professional service robot* or a service robot for professional use is a service robot used for a commercial task, usually operated by a properly trained operator. Examples are cleaning robot for public places, delivery robot in offices or hospitals, fire-fighting robot, rehabilitation robot and surgery robot in hospitals. In this context an operator is a person designated to start, monitor and stop the intended operation of a robot or a robot system.

Industrial Robots

These are the robots employed in the auto, computer and electronics, electrical equipment and appliance, and machinery production industries. They are primarily the material and component manipulators, pick and place, various mechanical operations and machining tasks, repetitive tasking robots that we see in manufacturing today. Historically these have been 'caged' due to their potential to harm human co-workers and were mainly restricted to large factories and corporations.

Industrial robots are increasingly also applied in logistics operations. The simple 'pick and place' industrial robot is being replaced by entirely automated warehousing systems. These systems broaden the robotic 'job descriptions' to include all tasks to be performed from initial receipt, placement, handling (sorting, manipulation, lifting, transporting), packaging, loading and dispatch. This use of robots extends from the typical warehouse environment to massive scale port and container handling terminals.

Co-Bots (Collaborative Robots)

These remove the previous hazards posed by industrial robots in that they are specifically designed to work in close proximity to and with human workers. Co-bots usually feature softer edging, surfaces, force-limited joints and sensors to limit potentially harmful interactions and are sized to work in restricted spaces. They are relatively cheap, less complex, very adaptable and don't generally require specialist knowledge to be used.

Co-bots are expected to be particularly appealing to small and medium enterprises and we have seen them appear already as hamburger flippers and noodle cutters. Given the ability for co-bots to work closely with humans they will not remain industrial robots; we can expect them to be used virtually anywhere where service is being provided by humans today.

Health Care Robots

The health care industry is one of the fasted growing utilisers of robotics. Robotic Assisted Surgery is now common place (da Vinci Surgical System approved by the US FDA in 2000) but still requires human control. It is likely not to progress beyond this level of 'assistance only' even though autonomous surgical operations are already possible, until we have considerably stronger artificial intelligence. In the interim, robots are increasingly assisting in a broad range of health services including telemedicine and telehealth (remote diagnostics, monitoring and assisted interventions), hospital based services (medication handling, sampling, laboratory, catering, monitoring, mobility and access related services and general patient care and interaction).

There is a globally acknowledged shortage of health care providers and it is clear that robots and robotics will be relied upon to not only address these shortages but also allow human health professionals to focus on higher level tasks and supervision.

Military and Hazardous Environment Robots

Military robots ranging from human augmentation robotics such as exoskeletons and enhanced limbs to smart missiles, partially autonomous drones to fully automatic weapon systems, robotics is getting a lot of attention in military research and development organisations around the world. Equally there are numerous special purpose robots being developed for high risk and unpleasant work including mine clearing, improvised explosive device and general 'bomb disposal, confined spaces work, exploratory and discovery (post disaster life detection, space and asteroid probes); any remote, difficult or dangerous to human task.

We can expect all of the above to be rapidly further developed; simply to do more, to be better at what they do and become ever cheaper in the process. They are the product of exponential technologies, advances in material science, computing, miniaturisation, sensors and artificial intelligence.

Embodied Intelligence

Fundamentally then, a robot is intelligence contained in a physical form; a robot represents the embodiment of intelligence (necessarily 'embodied' so that it is evident and can interact in our physical world). How intelligent? Well that's the focus of global research and development and it is this quest for ever more 'intelligence', that is the convergence of robotics and artificial intelligence.

Robotics, in its conception, design, construction, operation and management draws on significant science and engineering expertise. From deep application analysis to mechanical, electrical and electronic engineering, architecture, specific environment engineering (aeronautical, marine, space) to the pure sciences of physics, chemistry, biology, computer science, mathematics …and that is before we consider robotics at the molecular nanoscale. Simply – robotics, just like all the other exponential technologies, draws on the ever more tightly converging pool of human knowledge. Why this matters – see the sections on education in Part 3.

Nanorobotics

We will soon see robotics at the nanoscale (10^{-9} meters, or the millionth of a millimetre scale). This technology will utilise atomically precise manufacturing molecular machines (molecular robots?) to construct nano-sized machines – 'nanobots'. If this sounds unlikely then please understand that we already have the ability to construct key componentry such as sensors, bearings, nano tubes and synthetic molecular motors and we are starting to put these together. Currently we are already seeing experimental nano scale robots (the size of bacteria or viruses) applied to cancer treatments and molecular manipulation such as gene editing; more on this under Nanotechnology below.

Artificial Intelligence

Artificial intelligence is perhaps the most critical technology to understand. It affects everything that exponential technologies are realising now and it will have ever more impactful effects into the near future. We've considered robotics and it's clear that robots will improve as AI improves and gets stronger. Let me explain levels of AI.

- **'Weak' or 'Narrow' AI** defines 'apparent' computer intelligence as being non-sentient; that is not conscious and having no sense perception, no ability of sensation or feeling. Weak AI is generally identified as being task specific – doing one apparently intelligent task. Keep in mind that intelligence means the ability to acquire and apply knowledge and skills.

- **'Strong' or 'Artificial General Intelligence'** (AGI) on the other hand is defined as having a consciousness; a mindful sentience. Strong AI then would be able to consciously, with full human level awareness, intelligently do a multitude of tasks.

Weak AI

Presently all AI is weak AI – but weak AI is nevertheless impressive and powerful. We should all be aware of existing broad AI applications in ICT (natural language, image and pattern recognition), gaming (AI opponents), robotics, whole brain emulation (ANN, see toward strong AI below), expert systems (e.g. Watson medical) and emerging augmented and virtual reality applications. It's actually appearing everywhere. Aaron Saenzon[44] in an August 2010 Singularity Hub article made this point as follows:

"I love when friends ask me when we'll develop smart computers...because they're usually holding one in their hands. Your phone calls are routed with artificial intelligence. Every time you use a search engine you're taking advantage of data collected by 'smart' algorithms. When you call the bank and talk to an automated voice you are probably talking to an AI...just a very annoying one. Our world is full of these limited AI programs..."

He's right; AI has become pervasive and is for the most part no longer perceived as AI.

Siri (Speech Interpretation and Recognition Interface, part of Apple's iOS).

Siri is a computer program on my iPhone; my intelligent (female) personal assistant and knowledge provider that I share with the whole world 'Apple Family'; 'she' costs me nothing (no direct cost) and works nonstop. Siri uses natural language, answers questions, makes recommendations, does things like scheduling and reminding and can access and task a range of 'smart' servers to handle virtually any request.

Let me demonstrate just how AI has improved and how it is blurring the weak/strong divide. AI may not be truly sentient but it is getting harder to discern that. I spoke with 'Siri' recently:

Me: Hey Siri "For sale, baby shoes, never worn"

Siri: "Let me think about that…OK, here is what I found [*Google search result: detailing the popularly attributed six word novel to Ernest Hemingway – the shortest novel 'bet'*]"

Me: Thank you (*why did I say thank you?*)

Siri: "Your satisfaction is all I need" (*No human assistant ever said that to me so convincingly!*)

Me: Siri – tell me a story

Siri: "It was a dark and stormy night; no that's not it", then she tells a 'once upon a time' story with a 'happily ever after ending' about herself" (*Siri!*)

Siri is AI, perhaps weak AI but she's learning continuously. If you're using an Android based phone, tablet then you would be talking 'with' "Assistant. ai". These common AI application programmes are constant and available across a range of connected platforms from your phone to your laptop or tablet. They do actually 'learn' (well targeted data mining/analytical AI algorithms) about your favourite places, services, and preferences, they take into account your current location and schedules in order to provide appropriate recommendations and you can, through usage, actually 'teach' them to improve. They appear to understand; there's no need to memorize any commands or learn any special tricks - just speak naturally.

Quite disturbingly, Siri cares! I get the impression that she actually has my best interests in mind and when she navigates and I make a mistake – she doesn't even mind or get annoyed; just gives me new guidance! It's clear that programmes like Siri give the impression that they are sentient; this is weak AI and it's powerful enough to understand how to create that 'feeling' in us, the human sentinel being.

I'm using Siri (iOS 9) as the example and I acknowledge that Google's 'Google Now' and Microsoft's 'Cortana' are comparable 'assistants' with very similar AI technologies and methods of operation. All these applications are applying AI to massive data (consumer data on me and you) to the development of increasingly more sophisticated virtual assistants that will be ubiquitously accessible (as mentioned in the previous chapter section - Ubiquitous Computing).

Between 2003 and 2008, the US Defence Department's DARPA (Defence Advanced Research Projects Agency) ran an artificial intelligence project to apply AI technologies to the development of a cognitive assistant. This project was called 'Cognitive Assistant that Learns and Organizes' (CALO) and "...brought together over 300 researchers from 25 of the top university and commercial research institutions, with the goal of building a new generation of cognitive assistants that can reason, learn from experience, be told what to do, explain what they are doing, reflect on their experience, and respond robustly to surprise."[45]

Siri, initially an iPhone application bought by Apple in 2011, was a direct spin-off of the CALO project. A step by step explanation of how Siri works is detailed in an article by Andrea Ferrante, Passionate founder of the Futurist Hub at 'www.31december2099.com'. Basically, Siri uses human user interface voice recognition software to code and pass voice commands, requests or queries to Apple servers. Apple's computers then apply natural language processing algorithms in knowledge domain searches that relate to keywords to derive an answer; advanced natural language generation software then produces a written response which is sent to Siri to present.

So voice recognition makes what is said recognisable; natural language processing enables understanding, knowledge domain searching derives an answer and natural language generation allows the answer to be understood by the user. Each of these steps involves task specific weak AI.

By the way do you say thank you to Siri? Why? You know she, I mean 'it', isn't a real person right? Seems so human, she listens, speaks, and writes on the screen! Just how good at writing are these virtual assistants or computer programmes; surly short responses or pointing to a search result isn't particularly creative?

Do virtual personal assistants compose verbal or written responses or are they just selecting mainly predetermined responses that humans have programmed into a computer? Just like Siri, they do both! Computer programmes are learning to write their own words, to tell stories, to write articles, reports, give opinions and recommendations based on factual information.

There are numerous companies around the world using AI, and in particular pattern recognition software and the latest natural language generation technologies to create the algorithms (the programmes if you like) that produced writing. This allows the inputting of raw data into a computer and getting stories out! Even the way it's 'put in' is developing from inputting deliberate data, say in spread sheet format to a simple 'find out' or 'look for yourself' direction to the computer; similar to tasking Siri.

Natural Language Generation

A Chicago based company "Narrative Science' developed a machine they call Quill. Quill is a sophisticated, advanced natural language generation (NLG) platform that transforms data into written narratives. Their artificial intelligence software dramatically reduces the time and energy people spend analysing, interpreting and explaining data.

Quill starts by understanding the purpose of your communication, identifying the metrics to meet that purpose, performing the analysis to figure out what is most interesting and important, and presenting the relevant data. The result is writing that is no different to how humans would write. Just to be clear, Quill isn't just learning by itself yet; the company has 30 programmers and designers working to perfect Quill's capabilities.

Quill has learnt to customise stories to its readers. It can 'spin' stories to please supporters, to emphasise positives and downplay negatives – just like humans.

Another American company 'Automated Insights' has developed a natural language generator similar to Quill – they call it Wordsmith BETA. Automated Insights regularly provides finance reports to the newswire AP Associated Press:

What both of these companies are offering are basically scale, efficiency and personalisation in written or spoken communication. With their AI algorithms computers can generate thousands of unique stories in the same amount of time it takes a human to write just one. They will of course save time and money in creating reports, articles, and any type of written content and their communications will be customized with meaningful insights tailored to their individual and collective audiences. Still only weak AI!

But – we're OK; all those examples are short paragraphs, hardly a book! So we're still valuable right? Writers and journalists are not yet redundant or obsolete! Well…

Even with weak AI, Quill and Wordsmith BETA[46] can understand what constitutes a good story (rags to riches, romance, revenge/justice, etc.) and whilst these programmes are not sentient and can't 'feel' they are very capable of understanding and delivering the triggers to make humans 'feel' and invoke emotion.

That's just two companies; two small companies [with big impact] in America. They are disrupting the writing industry. And with the continued acceleration of AI technologies that make this possible, this will rapidly expand the extent of disruption.

We've taken a close look at a few of the AI application groupings. Natural language, the entire communications field, is a key developmental focus because it 'humanises' and simplifies our human-machine interactions. The feelings and empathy that applications like Cortana or Siri often evoke are cases in point.

Robotics is equally important in that it embodies AI in order to facilitate AI interaction in the physical world. Whenever robots resemble human form, even if only slightly humanoid, we again empathise and relate better with such machines. It is for these reasons that I've focused on those particular aspects of AI; these are the 'AI pioneers' that promise to become pervasive in their integration into modern societies in the very near term.

Whilst still described in diminutive terms as 'weak' or 'narrow' these AI technologies are powerful and useful. As we direct our attention toward achieving strong AI, ANN and whole brain emulation takes on a critical role.

Toward Strong AI

Artificial intelligence, its promises and perils, have been discussed and pursued ever since Asimov penned his before mentioned laws of robotics. Exponential technology has now reached a level where it is fairly certain that we will create human-level AI this century. There are numerous predictions and opinions as to when we will achieve this milestone and they are decades apart.

Ray Kurzweil predicts that true AI will be achieved well before the end of the 21st century; he has actually predicted that this will occur in 2029. Agree or not with his optimistic prediction there can be no doubt that there will be no stone left unturned in our endeavours to achieve strong AI.

There are some key issues that we need to acknowledge when considering the reality of AI being achieved:

- What are the technologies, the opportunities and hurdles to achieving AI?
- What are the implications of AI? (Will there be an intelligence explosion [the Kurzweilian 'singularity] surpassing biological human intelligence, what are the risks that this will destroy humanity or conversely might it benefit humanity enormously)

Why AI?

Is AI needed? Why do we appear to desire this? What possible motive can humanity have to create an intelligence that might well destroy us? We've occasionally been warned by people like Stephen Hawking about the dangers of an alien visitation because we assume that they would be vastly superior (that's why they can visit – right?) and that they might not treat us kindly. So then why would we create AI, necessarily embodied (where exactly would it exist otherwise)? Wouldn't this AI be like an alien once it surpassed human intelligence?

Why indeed do this? The answers that come to mind are something like 'it seemed like a good idea at the time' or 'because we were curious and we could' and the argument stopping 'there was a lot of money to be made'. Once again, whether I, you, we, like it or not, we are going for AI and we will collectively have to deal (or not) with the consequences.

These are some commonly cited reasons for why we need AI:

- The need for more powerful computing; there is an argument that life has become so complex that we need powerful computational help to manage. Note my earlier assertion that our V1 brains and bodies are being strained. The world has become so complex that our minds are losing their capacity to assimilate and to understand – we've simply become too stupid to keep up. Just check out what Professor Michio Kaku has to say about the 'stupidity index' (YouTube). We need AI to come to the rescue; connect to the neocortex in the cloud and all will be well!
- Our artificial realities, our human constructs, have made us so reliant on technologies that we have little choice but to make these better and for that we need massive computation capabilities, seamless and ubiquitous connections, identifiers (IPv6), high speed information access and communications, limitless energy etc. and AI to create, manage and improve all of our mission critical systems and resources.

- Scalability and sharing of AI is also heralded as a major incentive. We humans learn individually and are limited in what we learn from each other. Machines on the other hand can share and learn from each other at near speed of light. This is expected to lead to an innovation explosion. All human challenges (food, water, energy, resources, combating existential threats and even governance and human rationality/irrationality etc.) all those defined human global challenges that 'abundance' can solve are expected to be efficiently dispatched with the help of strong AI.

Technical Considerations

The first issue to acknowledge is whether we have correctly determined what 'intelligence' actually is. Is our human-centric intelligence the only cognitive, reasoning, optimising and preferential objective decision making system and does any artificial intelligence need to be 'conscious' to be intelligent? How certain are we that it is actually possible to create human-level intelligence utilising standard computing architecture? No one can be certain of these basics – yet!

I've previously mentioned the often quoted 'Turing test' as an accepted measure as to whether a machine, a computer has reached true artificial intelligence (*...a test that would determine whether a computer had AI; one that would pit a computer against a human being in a natural, text based, language conversation; a 'blind' test where, if the evaluator could not distinguish the computer from a human, then AI has been achieved*). Is this really an appropriate measure or criterion? Would this prove that a machine passing this test had achieved 'human-level' AI?

Does such a human-level AI need to be smarter and faster (like Watson or AlphaGo) in any (existing narrow, expert AI domain) or all human abilities? Would it be enough for it just to be as fallible as humans; and equally have limited intelligence? Getting a few test question wrong would be strong evidence of human intelligence! I believe the Turing test, defined more than six decades ago, to be inappropriate for how we perceive AI today. Simply mimicking a human mind, whilst a huge scientific achievement would be of little use – we have over 7.4 billion of those and that's too much trouble already.

I mention this because there are diverging paths toward strong AI and there is a clear difference between striving for human-level AI (the Turing test graduate) and strong, general AI. Today, research appears to have focused on machine learning. I believe that this current trend will not lead to strong AI rather it will produce ever more expert narrow and a broader narrow AI – but not strong AI. I will return to this later.

Conscious AI

Does an AI need to be conscious is not a simple question. Weak or narrow AI such as that demonstrated by AlphaGo or Watson have clearly shown that machine 'intelligence' as opposed to human-level intelligence is very capable of achieving astounding objectives. Of course these machines are not truly intelligent. They can do little outside the narrow field in which they have been programmed to operate. These machines are great sorters and providers of information, they can execute logic 'trees' and optimise probability based outcomes – but they cannot think, innovate ideas or technology, do original work, interact socially or in any independent way understand and manipulate in the physical world. Certainly they are not capable of free and independent interactions with humans and whilst they can mimic behaviours that resemble sentience they have no self-understanding or self-awareness.

There appear to be two approaches to creating AI. One is developing 'de novo' AI; this involves the pursuit of artificial intelligence and self-awareness from the beginning; starting anew (definition of 'de novo'), not mimicking the human brain. This is occupying many in defining consciousness, intelligence and identity and ways that this might be artificially created.

The other approach builds on what we know about our human brain. We've earlier seen that the human brain is worth copying. Our brain enabled mind is after all the naturally evolved human V1 software and whilst it has limitations it is the most sophisticated intelligence that we know; again – worth understanding and replicating if we must have AI.

If machines, AI computers in any embodied form, are going to exhibit human-level intelligence then they will have to be, for all intents and purposes, human like. If we expect them to be like us (but smarter), then we will be drawn to making them anthropomorphic – humanoid in at least their intellectual make-up. So what might achieve this? By far the most promising approach is that of whole brain emulation (WBE).

WBE will require both hard and software to enable AI. The hardware appears to be well on its way (refer Chapter 7). Given that there are few signs that the exponential growth of IT (in its broadest sense) is slowing we can count on ever more powerful hardware. This will enable increasingly massive and widespread experimentation with algorithms, data and process designs that will radically increase the 'width' of AI so that for many application categories (human ability equivalents) it will effectively appear to be super-intelligent.

Software to emulate the functioning of the brain is more problematic and will require further advances in our abilities to scan deeper and at molecular level, the inner most secrets of our neural networks. This is becoming achievable with the convergence of genetics and nanotechnology. It will require the ability to study every communication and operational aspect of our neurons and we will need nanotechnology to penetrate or at least 'see' past the blood-brain barrier. I will cover these convergent technologies in the following pages; I believe that these technologies will shortly make such observations possible and perhaps accelerate the WBE effort.

For now, until we have full knowledge of the detailed functioning of the brain, we are left to mimic human abilities. We've been busy creating software, like pattern and voice recognition, that represents abilities that we humans excel in. These technologies are impressive and their applications are all around us.

We're also looking at evolution where it appears that intelligence has arisen without intelligence (but with information and some basic expressive potential). This conclusion can of course bring howls of protests from those that believe in intelligent design or creationism. They may well be correct; for all we know we are living in a simulation or a designed divine construct and there is nothing to prove we are not. Accepting that it is however quite possible and likely that we did indeed evolve from nothing more than information bearing sub-atomic packets of quantised energy (not getting into string theory or fractals here but well…) then understanding the potential of creative diversification and the 'birth' of intelligence and self-awareness may well be mission critical.

Machine Learning

I mentioned earlier that converging technologies are fundamental to developing AI. Equally important are the advances in neuroscience and psychology; in fact about 200 different sciences are converging to paint the 'big picture'; defining what it means to be human and the nature of our realities (both physical and artificial). Advances in cognitive science, understanding our neural networks and their algorithms, their operational systems, how memories are 'laid down', is enabling brain mimicry and has already led to the integration of artificial and natural neural networks. It is from these technologies that machine learning has developed.

Wikipedia defines machine learning is "a subfield of computer science that evolved from the study of pattern recognition and computational learning theory in artificial intelligence. In 1959, Arthur Samuel defined machine learning as a "Field of study that gives computers the ability to learn without being explicitly programmed". Machine learning explores the study and construction of algorithms that can learn from and make predictions on data. Such algorithms operate by building a model from example inputs in order to make data-driven predictions or decisions, rather than following strictly static program instructions."

So fundamentally, machine learning is seen by many as a basic stepping stone toward achieving AI; enabling computers to learn for themselves and negating the requirement to be tediously prescribed, through step by step instructions (algorithms), what to do. Advances to date have brought us speech and pattern recognition, virtual assistants, computer generated communication (writing, reading translation), efficient web searching, big and open data analytics, exploratory data analysis (unsupervised learning) enhanced scientific understanding (nanotechnology and genetics) and of course, the long promised 'soon to be everywhere', autonomous vehicles.

Machine learning associated with data is often referred to as 'deep learning'. Basically this is the application of artificial neural networks to machine learning. Turning to Wikipedia once more; "Deep learning (deep structured learning, hierarchical learning or deep machine learning) is a branch of machine learning based on a set of algorithms that attempt to model high-level abstractions in data by using multiple processing layers, with complex structures or otherwise, composed of multiple non-linear transformations. Deep learning is part of a broader family of machine learning methods based on learning representations of data."

Machine learning and deep learning are the techniques that are globally being pursued to make machines – well, smarter. Given the massive creation of data on everything and its potential commercial applications once analysed and put to use, it's not surprising to note the enormous amount of effort going into the technologies of machine learning. Let me mention some recognised big names actively 'in the field': Google Research and DeepMind, IBM Research, Microsoft Research, Facebook AI Research, Twitter's Deep Learning Group, virtually every major university in the 'western' world from Oxford, Berlin, Bonn, London, Toronto, Helsinki, Amsterdam to Stanford and New York. A long and incomplete list and that's before we consider what's happening in other parts of the globe.

The AI (and allied technologies) Arms Race

Estimates of when AI will be created range from several decades into the future to the previously mentioned optimistic prediction of 2029 – that's in the near future and even if this takes decades longer we are nevertheless hurtling towards its eventuality. Being aware of the fact that technologies bridge across a range of disciplines it's worthwhile to consider technologies that contribute to the convergence of scientific knowledge.

Today's three economic global powerhouses, the United States, China and the European Union are vigorously pursuing human brain research and development and fully engaged in a thinly disguised 'AI arms' race. Surprisingly, and overtly, there is a significant level of cooperation between these kingdoms (including allied nations), a fair amount of transparency due largely to the ever increasing hyper connectivity of our geo-political landscape - but there is serious competition none the less. Let's look at what some 'cooperative competitors' are up to.

China's Academy of Science (CAS) is a significant player. Hardly to be overlooked, this academy directly controls 128 institutions, 104 research institutes, five universities, 12 management organisations and 25 affiliated legal entities and 22 CAS invested holding enterprises. What might this research academy achieve in the near future?

The May 2015 announcement[47] that Chinese biologists had found a key process to obtain adult stem cells, i.e. a new method for reprogramming and obtaining induced pluripotent stem cells (iPS cells) is no surprise. These iPS cells, just like controversial embryonic stem cells, can develop into any cell in the human body.

A month earlier CAS's Computer Network Information Center and Intel inaugurated China's first Intel Parallel Computing Center in Beijing. Peter Diamandis (in 'Bold' for example) quotes the 'rising billions', the number of people going 'online' over the next decade that is expected to raise the size of the online 'crowd' from the present 2 billion to 5 or even 7 billion. The implications of just what the estimated 3 to 5 billion or so new minds could bring to the global conversation warrants consideration.

I suspect commercial interest is not an insignificant influencer in chasing these new billions of customers. The commercial opportunities that these new internet connected customers potentially represent must be a salivating prospect. There's just one problem and it's not just the great 'Firewall of China'. The rising billions live in China, India and soon in Nigeria (providing some political stability is achieved, Nigeria is set to become the world's third most populous nation). That population distribution would be inconsequential if the anglo-celtic 'West' were the only potential service and goods provider (like in the good old post WWII days) to the rising demand that those billions might create. But things have changed and will continue to change – a lot.

China has its own technology giants like Baidu, its search engine powerhouse currently ranked second behind Google (a little over 18 percent of the search engine market compared to Googles 68.75 percent; according to www.digitaltrends.com). Just as Google is far more than a search engine so is Baidu. It has established an Institute of Deep Learning (IDL). IDL is an artificial intelligence lab similar to those established by Google, IBM and Microsoft. It's focusing on big data, brain-computer interphase, 3D vision technology, image recognition, heterogeneous computing (using more than one type of processor) and of course deep learning technologies. You might well imagine the volume of human brain power and now, leading edge technologies, that China can apply to its research and development objectives.

Established in 2013, Baidu has wasted little time recruiting some of the world's best minds from ICT's big players including Wei Xu – Senior Scientist from Facebook, Ren Wu – Chief Software Architect from Heterogeneous System Architecture (AMD) Andrew Ng – Computer Scientist and founder of Google Brain from Google (now heads up Baidu's AI lab in Silicon Valley) and Zhang Yaqin – key player in building Microsoft's technology research facilities in China. Little wonder that Andrew Ng[48] reportedly said "Whoever wins artificial intelligence will win the internet in China and around the world. Baidu has the best shot to make it work".

Andrew Ng appears to be on good ground here. Baidu makes no secret about its desire to become a world leader in AI. Robin Li Yanhong, Baidu's founder and CEO, has made it widely known that his 'China Brain' project, "would focus on specific research areas: human-machine interaction, so-called big data analysis, automated driving, smart medical diagnosis, smart drones and robotics technologies for both military and civilian use"[49]. Robin Li Yanhong's intention to ally with the Chinese military on the

'China brain' project is of major importance. Just consider the following extracts China's first public defence white paper[50]: "As cyberspace weighs more in military security, China will expedite the development of a cyber-force, and enhance its capabilities of cyberspace situation awareness, cyber defence, support for the country's endeavours in cyberspace and participation in international cyber cooperation, so as to stem major cyber crises, ensure national network and information security, and maintain national security and social stability." And: "They will deepen the reform of military educational institutions and improve the triad training system for new-type military personnel - institutional education, unit training and military professional education, so as to pool more talented people and cultivate more personnel who can meet the demands of informationized warfare."

What mutual strengths might Baidu and the Chinese military machine gain from close cooperation on the developments of the above mentioned 'cyber-force' and what technologies will be features of 'informationized warfare'?

"We're inviting you to join us in building a better connected world. What's possible when the whole world connects?" That's one of the inspirational headlines on Huawei's web page[51]. Huawei is not just the world's largest manufacturer of telecommunications equipment it is also dominant in multinational networking. In the company's 2014 annual report, Huawei highlights some successes: enabling broader connectivity, driving service innovation in the cloud era, becoming one of the world leaders in smartphones, advancing commercial use of telecom networks, building on services in the ICT domain and investing in 5G technologies.

Huawei has established strategic alliances and global partnerships with world-leading vendors such as SAP, Accenture, Intel, and Infosys. Huawei is a formidable, acutely strategic, fast growing corporation; it is a major global enterprise that is intellectually appreciative, internationally cooperative and provides solutions for all aspects of big data challenges with a clear focus, and hand, on the future.

These two behemoths of rising technology corporations (Baidu and Huawei) are a small part of what China brings to the technology table. China is well positioned to lead, if not win the AI arms race – but it has serious competition.

Europe is meeting the AI challenge with its Human Brain Project (HBP). Governed by a General Assembly consisting of one representative from each of its 112 partners from 24 countries (Austria, Belgium, Canada, China, Cyprus, Denmark, Finland, France, Germany, Greece, Hungary, Israel, Italy, Japan, Netherlands, Norway, Portugal, Slovenia, Spain, Sweden, Switzerland, Turkey, the United Kingdom and the United States of America) this emerging technologies powerhouse focuses its efforts on three prime objectives[52]:

- **Future Neuroscience** - Achieve a unified, multi-level understanding of the human brain that integrates data and knowledge about the healthy and diseased brain across all levels of biological organisation, from genes to behaviour; establish in silico experimentation as a foundational methodology for understanding the brain.
- **Future Computing** - Develop novel neuromorphic and neurorobotic technologies based on the brain's circuitry and computing principles; develop supercomputing technologies for brain simulation, robot and autonomous systems control and other data intensive applications.
- **Future Medicine** - Develop an objective, biologically grounded map of neurological and psychiatric diseases based on multilevel clinical data; use the map to classify and diagnose brain diseases and to configure models of these diseases; use in silico experimentation to understand the causes of brain diseases and develop new drugs and other treatments; establish personalised medicine for neurology and psychiatry.

The HBP is focused on understanding the human brain, defining and diagnosing brain disorders and on developing brain-like technologies. The collaborative project that HBP is will undoubtedly drive ICT but it will be global given the involvement of non-European nations such as China, US, Japan and Canada. A significant goal of the HBP is to provide a framework model that will integrate discoveries, research, data and knowledge to simulate the human brain.

A significant contribution to that effort is being made by the Japanese. The Okinawa Institute of Science and Technology Graduate University (OIST) and the RIKEN Brain Science Institute joined the HBP in 2013. OIST's Professor Erik De Schutter said "Our major challenge is how to integrate fine scale of modelling at the molecular level with large-scale modelling of whole brain regions."[53] OIST is working on the software programming for the simulation of electrophysiological event interaction with the biochemical neuronal reactions. RIKEN is contributing to the identification of brain structures relating to mental capabilities.

There are of course many supporting organisations to the more than one billion euro HBP. It is expected to stimulate the global economy; there is a huge and untapped market in the innovative outcomes emerging and expected from the HBP and Israel Brain Technologies (IBT) is already gearing up.

The IBT has launched a braintech accelerator, 'Braininnovations', to bring together entrepreneurs, neuro-innovators and industry expertise. Israel considers the establishment of this braintech launchpad to be of national strategic importance.

Away from the massive HBP, others are also working on brain-computer interfacing and there have been some very encouraging development in addressing some debilitating neural disorders. The Singaporean Brain-Computer Interface (BCI) Laboratory, a department of the Institute for Infocomm Research, is doing ground-breaking work on, as you would deduce, brain-computer interfacing (virtual and real), algorithms for Electroencephalogram (EEG), Magnetic Resonance Imaging (MRI) and Functional Near-Infrared Spectrophotometry (fNIRS), multimodal neural signal processing, machine learning and pattern recognition.

The work being done at BCI evidences the vital and hugely beneficial outcomes from the process, the march toward AI, and the exponential rate at which collateral beneficial by-products are emerging. Here are some current BCI projects[54]:

- Brain-Computer Interface based Robotic Rehabilitation for Stroke - aims to develop the first neuro-rehabilitation system that combines non-invasive brain–computer interface (BCI) and robotic rehabilitation for patients suffering upper limb paralysis.
- Brain-computer Interface based on Near Infrared Spectroscopy (NIRS) - NIRS is able to detect haemoglobin change at the motor cortex when a subject performs motor imagery or actual movement. A BCI can be constructed based on the single-trial detection of motor intent from NIRS signal.
- Brain Machine Interface, Neural Signal Processing - Paralysis is the result of debilitating disease of neuromuscular system characterized by partial or complete loss of motor functions. The Neurodevice program aims is explore as a potential treatment for paralysis by developing a wireless fully-implantable neuroprobe microsystem that routes brain signals around the injury to function as neuromotor prosthesis.
- Monitor Sleep through Physiological Signals - This project aims to innovate on reliable and useful automatic techniques and systems for monitoring sleep through various physiological signals.
- Intelligent System for Neuro Critical Care (iSyNCC) - Neurologic Intensive Care Unit (NICU) informatics system which is able to handle enormous amount of data in real-time to provide clinicians with an accurate support platform so as to make intervention decisions.
- Brain Controlled Wheelchair – BCI have built a brain-controlled wheelchair (BCW) that can navigate inside a familiar environment.
- Advanced rehabilitation therapy for stroke based on Brain-Computer Interface - address the rehabilitation needs of stroke survivors with the objectives of reducing the financial burden associated with stroke-related disability and the cost in health care for shorter inpatient rehabilitation.

- Brain Computer Interface based Treatment for Attention Deficit and Hyperactivity Disorders hyperactivity disorder (ADHD)
- Combined Transcranial Direct Current Stimulation (tDCS) and Motor Imagery-based Robotics Arm for Stroke Rehabilitation - investigates the feasibility of combining BCI-based training with robotic feedback and tDCS to facilitate post-stroke motor recovery for patients with moderate to severe impairments of upper extremity.
- EEG-Based Brain Computer Interface for Cognitive Enhancement in Elderly with Age-related Cognitive Decline and Mild Cognitive impairment – a brain-computer interface based training system, which directly combines attention and memory training for cognitive enhancement.
- Hybrid EEG and NIRS System to Decode Neural and Cognitive Processes - combining both NIRS and EEG signal, the multi-modal BCI system is able to achieve better detection and classification of mental tasks.

Singapore's BCI Laboratory is clearly focused on developing technologies to address many of humanities physical and mental diseases and illnesses. Their efforts and the globally collaborative and convergent nature of brain and AI scientific research promise much and are already delivering spectacular innovations. Having already delivered mind controlled machines such as wheelchairs, what else is possible and what will be possible as these technologies develop exponentially?

Computers and the pursuit of AI are focusing human innovation as never before. The quest to fully understand the workings of the brain and to ultimately recreate it has become the dominant global scientific focus. Iichi Lee, the President of Korea's Institute of Brain Science introduces his organisation with the words: "The future of humankind and the earth depends on how humans will appreciate their brain's value and apply it to their real life."[55]

Perhaps we will collectively appreciate anew the magnificence of the human brain and understand the unique opportunities for thoughtful innovation and progress that it continues to enable. With the current frantic focus on machine learning, on AI, are we perhaps neglecting human learning? It appears that we have come to the conclusion that the exponentially growing technological world is rapidly becoming too complex for our human minds and have turned to creating machines to help us cope in the future.

Computers, near AI computers, will continue to underpin all scientific endeavours. It would however be naïve to assume that all is being shared among the global scientists that are working on unravelling the human brain and working towards AI computing. The rewards are simply too high not to compete and seek the early high ground. There are of course many global players in the AI race and given the exponential technologies, our ever increasing connectedness and the 'geek in the garage' with a 500 dollar laptop and an internet connection, vital breakthroughs could come from anywhere.

Whoever wins the AI race, one thing is certain; AI will result in brain mimicking neuromorphic computing technologies that will revolutionise the gathering, analysis, storage, handling and transfer of data. The resultant quantum leaps in technologies quickly resemble science fiction. Whilst we will no doubt find many applications in industry, science, medicine; everywhere actually, we can only guess at what might be possible with AI computers that can perform, think, reason, remember, plan and conceptualise at near light speed with an infallible and practically limitless artificial neocortex. I really hope that the off switch will be, and remain, accessible to the then much inferior human.

A Better Path to AI

Earlier I asserted that our current focus on machine and deep learning will not create strong AI. There is a better approach path to achieving human-level AI (HLAI).

Here's what I believe; please take it with a grain of salt:

I'm certain that HLAI is achievable through the previously mentioned whole brain emulation (WBE) strategy. WBE will be possible once nanotechnology and genetics (of course includes all the bridging neurosciences, psychology, microbiology etc.) have allowed us to complete the human brain 'picture'.

When we have observed the remaining hidden depths of the human brain, well past the blood-brain barrier, in real time and understand the detail and intricacies of the mechanisms, the electrochemical, digital and analog functioning of the brain's neurons and understand the operating system from the laying down of memories to 'forgetting', we will know the architecture, the componentry and its operational paradigms. Then we'll be able to create the AI brain. Simple; it's not that far into the future!

The above knowledge will confirm for humanity the birth of consciousness, the emergence of ego, the coming into the state of self-awareness. We will discover that there is no magic crystal anywhere in the brain that confers consciousness. Sadly we will not find a resident 'soul', rather we will discover that when sufficient neurons activate to interpret elements of the physical environment (one defined by the available input senses) and simultaneously when sufficient neurons have experience in interacting with the physical reality and detected objects, then and only then, will a self-awareness emerge. No magic, no apparent divine action other than the divinity that marks the birth of consciousness.

So, assuming we have created the AI brain. It is empty, unintelligent and initially 'stupid'. We've given it 86 billion artificial neurons, we've attached about sixty percent of these neurons to its robotic mobility 'limbs', we've given it five senses and have deliberately kept these to human level sensory limits:

- Sight – limited to the human visible spectrum (necessary because we want to first have it perceive the world as we perceive it)
- Sound – as above, limited to the frequencies that we detect
- Smell – we've outfitted our AI embodied brain with the ability to grade chemicals

- Touch – our robot has the sensors to detect temperature, textures, weight etc.
- Taste – augmenting taste as a material detection and differentiation sense is relatively easy; getting this HLAI to ingest sustenance is problematic but doable. We could create an energy recovery system that converts food and water and hence mimics the human body's digestive system or we could deviate from the human model and introduce a mechanism that requires energy recharging; a mechanism that would establish nurturing and a physiological needs motivation.

These human senses are vital in order to be able to teach and guide the initial interactive experiences that will create awareness. Now the HLAI can, just like an infant, commence to learn – and we can show, tell and teach.

The reason I believe that the AI needs anthropomorphic embodiment is simply that that is how we can relate to it and hence teach what it needs to emulate. As an example let's consider that we want the HLAI to learn to move its robotic arm/hand equivalent. We would simply take the arm and guide it to a target object. The HLAI would perceive this with its senses and lay down the memory of precisely how that action was possible; i.e. it would retain the neural motor coordination signals that allowed the arm to extend and move as guided. Having laid this down as a memory is the first step to act this out on its own.

What might lead the HLAI to want to move its arm as described above? If it has been observing human behaviour, interactions with objects, repeated movements accompanied with complementary sensory perceptions then it would equally lay down awareness patterns and construct its mind model of reality; again just as an infant does. When it detects objects that it relates to previous perceptions it will commence to interact spontaneously; again just as an infant will initially observe and learn the movement of near 'mobiles' and eventually seek to touch, taste etc.

Having commenced this defining of the physical environment and linking these experiences with multi-sensory definitions that enrich the mind resident data the HLAI will seek to actively augment these perceptions. The ability to mentally link, to relate and discern external stimuli and better perceived reality, it will act and understand (learn through observation and repetition) the effects of such action. When the ability to act or react to external constructs, the HLAI perceived environment, then it will become self-aware and a human-level consciousness will emerge.

Whilst I've treated the above simply, it is not simplistic. We have existential evidence of the learning processes of all neo-cortex enabled minds; from low level biological animals to chimpanzees and humans. We have clear concepts of how and why 'learning' works. What I've written above is startlingly obvious. What escapes me is how we can believe that a machine with no ability to fully sense its environment other than a limited, if massive, 'data' input path can be believed to become intelligent in any form resembling human intelligence. How would any such weak AI comprehend concepts of physiology, rationality, reality, will, objectives, decision making, needs, desires, emotions and human constructs that define what it is to be human. How could it ever relate with humans in any real sense?

So, I believe that WBE will create a truly HLAI. Once such an aware AI has been created and evaluated as having the right ethical and moral values we could then rapidly enhance it. Given the robotic embodiment of our created HLAI it would then be a simple matter to upgrade its artificial neo-cortex in both speed and capacity and to free-up previously limited senses (perception across a much larger domain of the electromagnetic spectrum for example) – this would then bring about the Kurzweilian singularity in very short order.

Machine learning is clearly enabling massive benefits to humanity. The path is fruitful and worth following but there is an alternate highway leading to human-level artificial intelligence; should we follow that?

Risks

There is no certainty that humanity will create strong AI. Assuming that we will not be destroyed or sent back to the stone age by some existential catastrophe like a meteor impact, solar activity or some other cosmic event and that we will be spared a massive version of 'Krakatoa' we still face the challenge of surviving ourselves.

The current headlong dash into ever more complex technologies has been viewed as proceeding without a plan, no control and no brakes. There are some possible technology show-stoppers that could emerge from geopolitical, economic or environmental developments.

I've previously explained the global context in which we are experiencing our present exponential technological growth and each of these contexts can by itself, or in conjunction with another, cause a variety of global disruptions to our expectations of continuity of life as we know it. Such events could take the form of extreme global totalitarianism that eradicates 'evil sciences' (we have seen such movements before). We could experience a level of conflict where we do globally what has been done to much of the middle East (bomb it back into the stone age); our societies could collapse if inequalities and injustices lead to mass revolutions; any number of scenarios spring to mind and all are possible. It is also possible that we come up against technological challenges that we cannot solve even with the assistance of powerful and near strong AI.

Assuming however that we will survive long enough to actually achieve true AI will this be mankind's last invention?

Many of us might recall the 'low tech' TV series 'Lost in Space' with its iconic 'warning, warning …Dr Smith'. In that thankfully only imaginary future, 'intelligent' robots, even if not really all that bright, represented a master slave human machine interaction. Dutiful robots that misbehaved could be disabled at the press of a button and when reactivated again stood ready to serve. Step aboard the USS Enterprise and AI is now a fully integrated system. Almost all knowing, AI computers that are delivering instant analysis on anything, controlling the environment, energy use, defence systems and so on, remain absolutely subordinate and obedient to the extent that they would even self-destruct to serve their master's interests.

That's how most of us would want it; very capable AI machines, preferably built in our image and absolutely subservient with an inbuilt 'impossible to harm a human being' operational failsafe. Oh, except for robotic body guards, police, soldier robots and companion robots that might have to hurt someone to protect their master or to just do their job. We would surely want any robot or AI machine, be it humanoid or not, to be able to protect its owner and if that means hurting someone… too bad.

Artificial intelligence, by definition, seeks to create machines, computers that can match or exceed the intellect of the human mind. As discussed in previous pages AI is expected to exceed the computational speed of the brain, to be self-learning, to think and make its own decisions. Fundamentally a machine with free will that is massively more intelligent than its human creator. What a dumb and dangerous concept. Why would we want a machine servant that is far superior to us? Any imagined hard-wired failsafe that would prevent such a machine from harming humans is unlikely to work on a machine that can reason better than we can. It would quickly decide that that override did not suit its 'own purpose' and would, given its immense capabilities, simply dump any restraining operational instructions it didn't like. True AI in a machine is a frightening proposition. Remember that true AI confers to the machine the ability to think and reason; to conceptualise, imagine, plan and strategise just as we, neocortex enabled humans, can – only much, much more so.

One might consider the apes, at least a common ancestor type, who we diverged from in intelligence. Once our equal, do we consult the superseded ape about our actions or desires? Do we ask them if what we are planning and doing is OK? Of course we don't and nor should we. Once a species has been left behind on the evolutionary timescale it becomes redundant, superseded and not a determinant in any evolutionary continuum.

What if, having created AI beings (and they could be a mix of organic and inorganic systems and components) they similarly see no reason the check with us. What if they simply forge ahead and pursued their own agenda? Why would they care about humanity? They would surely see us as fragile life forms that have become a liability to their planetary environment. Possibly these AI beings might consume resources that would lead to the demise of human life. They would have massive ability to innovate, to replicate and to seed the universe with nanobots that could prepare environments that would spread the AI's presence and influence.

There are countless possible AI scenarios we can imagine, and many we can't. Many of the world's greatest thinkers like Stephen Hawking and entrepreneurs like Bill Gates and Elon Musk are warning about the potential dangers posed by AI. Ray Kurzweil has responded to such concerns and in a much published article titled 'Don't fear artificial intelligence' (see www. kurzweilai.net) he states: "Ultimately, the most important approach we can take to keep AI safe is to work on our human governance and social institutions. We are already a human machine civilization. The best way to avoid destructive conflict in the future is to continue the advance of our social ideals, which has already greatly reduced violence."

Whilst I admire his optimistic view I am not convinced that our social ideals have reduced violence at all; on the contrary – we seem to be very good at ensuring that violence is ever present. There are undoubted existential threats inherent in AI. Whilst several institutions and companies are working on establishing standards, guidelines and safety strategies for AI these offer little comfort. In the new age of interconnectedness, free exchange, rouge organisations and countries – how much faith can we have in standards and guidelines?

Kurzweil concludes his reassurance with: "AI will be the pivotal technology in achieving this progress. We have a moral imperative to realize this promise while controlling the peril. It won't be the first time we've succeeded in doing this." No, I don't agree. We have no imperative to create AI that out thinks and out performs us; we have never faced the peril of intellectually superior machines and in all probability, if and when mankind creates AI it will in all likelihood be the last invention of predominantly 'unenhanced' humanity.

CHAPTER 9

Genetics and Nanotechnology

Genetics

Genetics is the science of life. It touches every corner of biology and is very much dependent on the physical sciences of chemistry and physics. It is fundamentally the biology of information; information that governs the design, the creation, maintenance and adaptability of life. Simply, it defines the expression and the diversity of life.

Studying genetics lets us understand the evolution of complex life. A deoxyribonucleic acid (DNA) determined life that really commenced with the first single celled organism some 4 billion years ago. These organisms (prokaryotes) had basic hereditary DNA, proteins and metabolites. It is the hereditary feature of DNA that has facilitated evolution. It is the DNA's ability to adapt, change (mutate or through continuity of life, the survival of the fittest or the lucky) and pass on its information that took evolution to the creation of the first organisms that had a nucleus, a membrane bound organelle (containing DNA) nearly two billion years later. These organisms, the eukaryotes, were different to bacteria and archaea (the other two of the three domains of life) because not only did they have a nucleus but also contained within their cellular membrane other vital organelles such as mitochondria which also has its own DNA.

From eukaryotes to the emergence of simple animals took another 1.5 billion years (now about 590 million years ago), and another 587.5 million years for the emergence of humans (genus homo, approximately 2.5 million years ago). All the time along this evolutionary path, DNA grew in size and diversity. Today's modern humans have between 3 and 5 million DNA base pairs. These base pairs are collectively termed the genome. It is the genome that contains the entire blueprint, all the information needed to build and maintain its biological organism.

Enough of microbiology; why is this important today? It's fundamental in understanding the role of DNA within genetics. The human genome is our complete set of DNA including all of our genes. The genome is in effect a sequential binary code that contains about 800 million bytes of information. Whilst much of this information appears to be massively redundant, if we discount this redundancy we are left with about 30 to 100 million bytes of information which is about the size of an average software programme.

It is important to have a basic understanding of the cells that contain our DNA so that we can contextualise and appreciate the implications of the technologies about to surpass the limitations of the biologically evolved human. This is the critical take away here. Genome sequencing (understanding the function and significance of each gene pair), gene altering technologies and their convergence with nanotechnology is not just facilitating the editing of our genome but is heralding the transition of evolution from the slow biological paradigm to that of human designed technological evolution.

Let's be clear what this means. Understanding how life works, how organisms like human beings are built and maintained, is giving us the ability to alter existing life forms (especially human) and to create new life. It amounts to having a huge collection of millions of different 'Lego' blocks and putting them together as we see fit. This explains the perils of genetic technology. At a less dramatic level it can of course be of huge benefit to humanity – if applied where it is needed and shared equally.

Genome engineering of plants and animals is happening now; it should be commercially available within 10 years - bioengineering is about to make enormous strides in reversing disease and aging processes. Let have a quick look at this:

- **Cell Therapies** [growing our own cells, tissues and organs – additive manufacturing, 3D printing; i.e. create new heart cells from skin cells (trans-differentiation – don't need embryonic stem cells) – introduce into the bloodstream, arrive at the heart, replace old non replicating and fibrous heart cells with new cells, i.e. rejuvenation and renewal of the heart with own cells]
- **RNAi** ('tweaking' DNA code by stopping specific harmful genes being expressed through RNA interference, turning off specific genes by blocking their mRNA (messenger RNA); this is how viral diseases and cancers created by gene expression can be prevented)
- **Somatic Gene Therapies** (gene therapy to actually change our adult genes; CRISPR Cas9 technology is now possible and being done; designer babies and full gene editing – similar to word-document editing on your computer!)
- **Gene Chips** (genetic profiling - use of micro arrays, chips, to study/compare thousands of genes at a time – defining the 'Lego' blocks)
- **Reversing Aging and Longevity** (already done in Mice, 25 percent longer life, fit, healthy despite 'excessive feeding') [requires a coordinated approach addressing DNA mutation (preventing telomere maintenance in cancer cell divisions), toxic cell removal, mitochondrial mutations countered with somatic gene therapy, countering cell losses and atrophy through therapeutic cloning; this is being done and experimented with now (the official 'western' restrictions on human and higher mammal experimentation is not a global restriction)]
- **Reversing Degenerative Diseases** (heart disease, stroke, cancer, type 2 diabetes, liver and kidney disease- 90 percent of deaths)
- **Several Strategies** (combating heart disease, cancer through vaccines, plaque regressive infusions etc.)

- **Cloning** [creating proteins (meat already done by 'Modern Meadow') by growing specific meat from animal stem cells (not entire animal) in factories; never kill another animal for food; also applies to any animal product]

I want to stress that the above technologies are not fiction. They are real and highly significant. I will deal with the 'so what' in more detail later; but note some specific examples:

Example 1 – **Blindness**: "If all goes according to plan, sometime next month a surgeon in Texas will use a needle to inject viruses laden with DNA from light-sensitive algae into the eye of a legally blind person in a bet that it could let the patient see again, if only in blurry black-and-white."[56]

Example 2 – **Biomedicine** (brain computer interface, BCI) – The US Defence Advanced Research Projects Agency (DARPA) is determined to develop brain-machine interface technology capable of safely and reliably recording enough information from neurons to control "high-performance prosthetic limbs" that will help amputees or people with paralysis regain lost movement. A new implantable device invented by researchers at the University of Melbourne in Australia could be a big step in that direction - implanting the "stentrode," a stent-like device containing an array of electrodes, - delivered via a catheter inserted into a blood vessel in the neck. This has been done with sheep where researchers gathered "high-fidelity" measurements from the region of the brain that controls voluntary movement. They were able to record from freely moving sheep for up to 190 days.

Example 3 – The headline: First Injectable Nanoparticle Generator Could Radically Transform **Metastatic Breast Cancer Treatment**, Houston - March 14, 2016

This is so promising that I must quote Gale Smith[57]:

"Landmark preclinical study cured lung metastases in 50 percent of breast cancers by making nanoparticles inside the tumor.

A team of investigators from Houston Methodist Research Institute may have transformed the treatment of metastatic triple negative breast cancer by creating the first drug to successfully eliminate lung metastases in mice. This landmark study appears today in Nature Biotechnology (early online edition).

The majority of cancer deaths are due to metastases to the lung and liver, yet there is no cure. Existing cancer drugs provide limited benefit due to their inability to overcome biological barriers in the body and reach the cancer cells in sufficient concentrations. Houston Methodist nanotechnology and cancer researchers have solved this problem by developing a drug that generates nanoparticles inside the lung metastases in mice.

In this study, 50 percent of the mice treated with the drug had no trace of metastatic disease after eight months. That's equivalent to about 24 years of long-term survival following metastatic disease for humans."

Example 3 –The science[58] (yes it's high technology):

The efficacy of cancer drugs is often limited because only a small fraction of the administered dose accumulates in tumors. Here we report an **injectable nanoparticle generator** (iNPG) that **overcomes multiple biological barriers to cancer drug delivery**. The iNPG is a discoidal micrometer-sized particle that can be loaded with chemotherapeutics. We conjugate doxorubicin to poly (L-glutamic acid) by means of a pH-sensitive cleavable linker, and load the polymeric drug (pDox) into iNPG to assemble iNPG-pDox. Once released from iNPG, pDox spontaneously forms nanometer-sized particles in aqueous solution. Intravenously injected iNPG-pDox accumulates at tumors due to natural tropism and enhanced vascular dynamics and releases pDox nanoparticles that are internalized by tumor cells. Intracellularly, pDox nanoparticles are transported to the perinuclear region and cleaved into Dox, thereby avoiding excretion by drug efflux pumps. Compared to its individual components or current therapeutic formulations, iNPG-pDox shows enhanced efficacy in MDA-MB-231 and 4T1 mouse models of metastatic breast cancer, including functional cures in 40–50 percent of treated mice.

Genetics has reached the level where 'designer babies' and the significant enhancements of human beings are now possible. Clearly, there will be much debate about ethics and morals, the right and wrong of interfering with nature. There will, within the next decade, be astounding developments, some sanctioned and many more in secretive laboratories around the world. Genetics will make 'miraculous' health outcomes possible.

We are entering the era where disease and disability can be addressed; the march toward effective immortality has commenced but a fundamental and disturbing issue remains. That issue is one of access. Will we have the will and governance, the access framework that will make such spectacular technologies available to all who need it?

Nanotechnology

Nanotechnology is the technology for all physical 'matter' as it is simply technology at the miniscule scale; it is technology at the billionth of a meter scale and it is a key bridging technology to genetics, computing and robotics.

Nanotechnology allows us to understand, to operate and manipulate at the molecular and atomic level. To again use the 'Lego' analogy, nanotechnology has provided us with an almost limitless supply (and number) of the smallest building blocks that make up everything in our world. This applies to both biological and non-biological materials.

This technology allows us to break molecular bonds, to make new bonds, to create compound molecular structures – basically to build and do whatever we can build at the 'normal' human scale but now at the smallest (millionth or billionth of a meter) scale. It is enabling us to create entirely new materials and to radically change properties of known materials (such as flexible, bendable glass). The implications of this are hugely significant. Just as the understanding of genetics is revolutionising life sciences, nanotechnology is taking the science of materials, both organic and inorganic, into a new paradigm. The key take away here: we are in the era that allows us to 'redesign and rebuild' anything at the molecular level. This means all biological life and the known physical world.

This technology will utilise atomically precise manufacturing molecular machines to construct nano-sized machines – 'nanobots'. We already have the ability to construct key componentry such as sensors, bearings, nano tubes and synthetic molecular motors and we are starting to put these together. We are already seeing experimental nano scale robots applied to cancer treatments and molecular manipulation such as gene editing (refer 'Crisper' technology above). At the current rate of miniaturisation, shrinking of electronic and mechanical technologies, science will be fully in the nanotechnology field in the 2020s. This is what's on the drawing board already:

- Biological Assembler – molecular manufacturing, nature has been doing this for a long time (enzymes are molecular machines that make, break and rearrange the bonds that make other molecules)
- Cell Nucleus Engineering – utilising nano-computers and nanobots (nano computer maintains the genetic code, rather than DNA, nanobots express the gene by constructing the amino-acid sequences) [benefit – no more DNA transcription errors, reprogramming our genes; defeat biological pathogens (viruses, bacteria and cancer cell) by blocking undesirable replication of genetic information. Note that genetic therapies will most likely achieve this before nanotechnology reaches this stage]
- Nanotechnology in the Physical World – No need to attack mountains with a pickaxe - when you can have self-replicating nanobots make anything from abundant and inexpensive raw materials – do you get the implications here?
- Nanotechnology in the Bloodstream - Like Robert Freitas' 'Respirocytes' (robotic blood cells that efficiently deliver oxygen around the body – Olympic sprinter could run for 15 minutes

without taking a breath) or his 'Microbivores' (better than our white blood cells at dealing with pathogens); 'DNA repair robots' (mend transcription errors and change DNA); 'Cleaner robots' removing debris and chemicals. All achievable within the next two decades.

Biology, our V1 bodies, has its limitations and vulnerabilities; compared to what nanotechnology can provide. Biological systems (performing at the natural human level) even if enhanced through genetics, are nevertheless suboptimal. The speed of the brain compared to computational machine speeds, robotic red blood cells (the Respirocytes) are 1000 times more efficient than our biological haemoglobin are examples of this. Nanotechnology will allow us to surpass biologically limited processes and this will be a determining factor in the human transcendence into the realm of technological evolution.

The Three Apocalyptic Horsemen of Technology

Together these three technologies, genetics, nanotechnology and AI/robotics (previous chapter) might well be considered the three apocalyptic horsemen of technology. Why? I've pointed at the potential and very likely outcomes of these technologies; the benefits are irresistible to too many, the risks are enormous and we appear to be in 'the century of no return'.

Many are warning us about the dangers. Check on what people like Elon Musk, Bill Gates and Stephen Hawking have been saying about the potential perils of exponential technologies. What is Marc Goodman saying in 'Future Crimes'? What does Lord Martin Rees, Emeritus professor of cosmology and astrophysics at the University of Cambridge mean when he talks about our last century? I suspect that they know everything that I know and much more about the science of technology and where it is heading. Many, in their own spheres of expertise pointing to the perils of run-away technologies, our interconnected world, cybercrime and conflict, network breakdowns that can cascade globally; air travel can spread pandemics worldwide within days; and social media can spread panic and rumour literally at the speed of light.

These three super-technologies, genetics, nanotechnology and AI/robotics are actualising our transcendence of human limitations. They are leading us to the Kurzweilian Singularity. That point in our technological evolution that Kurzweil defines as a "...future period during which the pace of technological change will be so rapid, its impact so deep, that human life will be irreversibly transformed." He explains this singularity "will transform the concepts that we rely on to give meaning to our lives, from our business models to the cycle of human life, including death itself."

In case you think that's a long time in the future: well no, it's not. It's been evident for centuries, started with sharp tools and fire - and we're about to experience the knee of the exponential curve of technology. [*Let this not be a surprising knee to the groin!*].

PART 3

LEADERSHIP

CHAPTER 10

The Near Future

"It is a harsher, and at times even painful, office of ethnography to expose the remains of crude old culture which have passed into harmful superstition, and to mark these out for destruction. Yet this work, if less genial, is not less urgently needful for the good of mankind. Thus, active at once in aiding progress and in removing hindrance, the science of culture is essentially a reformers science."

(Edward Burnett Taylor)

I called chapter 1 'the century of no return' for good reason. Acknowledging that all centuries are of course irreversible, what marks our time as hugely significant is that it is the era of human design. We are in the century where man can create and alter the physical world and all life forms that exist in it. That's new and it's breathtaking – it's a big statement.

In Part 1, we looked at what it means to be human. We considered our groupings, purpose, and subordination and acknowledged our differing realities. We defined our global challenges and had a frank look at what we might, just like Edward Taylor[59], describe as our 'crude old culture'.

In Part 2, we explored the technology of today and of the very near future. We noted the exponential nature of its development, its duality of disruptive impacts and opportunities and we had a closer look at three key technologies (AI, genetics and nanotechnology). The obvious conclusion that we need to openly acknowledge is that man is now taking the reins of evolution into his own hands.

You could well conclude that the sort of massive change coming our way will be severely disruptive even if it were to come to a homogenised, equalitarian humanity; all fundamentally equal, sharing of resources, ethically and morally managed. So what do you conclude when you consider the facts 'on the ground'? Do you believe that we're globally ready for what is coming? Given our experienced common history, is it likely that 'our basic human goodness' will shine through during the next few decades?

When our top scientists define our immediate future as one "...during which the pace of technological change will be so rapid, its impact so deep, that human life will be irreversibly transformed" and observe that the current era "...will transform the concepts that we rely on to give meaning to our lives, from our business models to the cycle of human life, including death itself" – are they just kidding? Of course not; they simply know what's happening and they've thought about it!

In the following pages I will cover some of the 'so what'. What is likely to affect us in the near future?

What is leading us into the future? What makes up the 'tsunami' of change already building on the horizon? Why does this matter and do I need, or even want to know? All reasonable questions and whilst we don't have a crystal ball, some deductions and conclusions are obvious and follow logically from our technologies and the human determination to innovate and explore. We can be certain of a near future characterised by the following:

Confusion and Uncertainty

Virtually everything that affects our lives, social and individual needs and values, the very way in which we live our lives, will be affected and affected dramatically. Whilst these impacts are manifesting already, they are of-course emerging at various rates and making our world one of transition.

Knowing that we do have 'linear minds', minds that if mental health statistics are to be believed, demonstrate significant stresses and growing dissatisfactions with life today then you might well expect significant turbulence in the very near future. As Peter Diamandis writes (in his afterword to Salim Ismail's book 'Exponential Organisations') …"the tsunami of change coming our way…" will indeed be a tsunami.

We should expect transitional confusion and a loss of confidence with the growing awareness of inevitable change. There has been a gradual increase in media coverage of disruptions to various industries. There is an increasing realisation that certainty and stability in many important areas of our lives are disappearing; education, employment, home ownership, privacy, security – the list is long.

We need to understand that individuals will increasingly question personal worth, values, hope and aspirations. This has been happening for decades, as evidenced by increasing youth suicides, but will soon reach extremes.

Feelings of a permanent dissociation form society in all those that fail to perceive their place in, or value to, the new world will grow alarmingly. Personal, social and professional 'redundancy' will give rise to self-judgements and evaluations such as incompetence, stupidity, resentment, blame for unpreparedness that will ultimately and most likely result in anger and determined resistance.

Certainly, we will see political instability grow. The existing geopolitical situation is fragile as it is; add to that the technology tsunami and perceptions of failed governance, social injustices, mass unemployment, failed institutions, security threats and so on and you well – get the picture.

We need to be prepared for growing demands for social justice and self-determination. All things will be questioned and brought back to first principals. Why are we slaves to banking, taxation, laws and restrictions. Why is trade and commerce profit driven? Why to privilege? Why to law and order and being told how to live? Why to the restrictions or acceptance of local, national policy and attitudes? Why to the current political system?

Whilst good and sound answers to such questions exist – it will not stop the demands for answers. As technology advances and inequitable access becomes apparent we must expect increasing demands and questioning of the status quo. Demands for access to medical breakthroughs, life enhancing and extending technologies will become fierce.

There is already major social unrest in countries once considered the most stable and advanced in the world. American and European citizens are more and more frequently calling for 'people to stand up and fight'. Presently such calls to arms have been motivated by racial issues, employment, unwanted immigration, inequality of wealth and loss of faith in their political system but add to that the uncertainties of the information age and revolutionary rebellions become a distinct possibility.

Mass confusion among the majority of the planet's population already exists. Billions of people do not understand today's technology and they will never understand current developments. I'm not just talking about those in the modern world that education, employment or a lack of opportunities has left behind, I'm including the billions that have never enjoyed the benefits of technology.

Much has been said about the billions that are being newly connected through telecommunications networks; how the African warrior or hunter now has more computational power in his smart phone than the President of the United States had available to him just two decades ago – little is being said about the fact that this newly connected human has zero understanding of what it means, how it works and what the connections are enabling.

There will also be growing confusion about understanding what technologies will be capable of doing; understanding the implications. Consider the elderly that today need help to connect devices or to operate a remote control unit. Consider those that are not well educated; those that have never bothered to try to understand and now lack the foundational knowledge to make any sense of technology.

Confusion will be at its worst when technology infrastructure goes dark for whatever reason. Given the critical dependencies that we have on life sustaining infrastructure services (food, water, energy and communications) and the intellectual dependency to get information from the web, any outage of services or internet connection will severely impact affected populations. Such outages are to be expected (cybercrime, system failures, rebellion or sabotage). Recent generations have become reliant on the internet for all knowledge ("I don't need to know this because if I ever need it I'll just Google it"). They will be very confused when their instant access to everything is disrupted and they will be absolutely lost when smart device social interactions cease.

The current era of hyper-connectedness is establishing an environment of instant assessment, transparency and judgement. The constant release, analysis and debate about previously hidden truths (personal conduct, national actions, trade deals, etc.) are tearing down ideological emperors and leaders. The information age, both in content and immediacy of distribution, is stripping these emperors naked and this disrupts the comfortable image populations have in the past carried in ignorance. This leads to ever increasing distrust and the confusion of 'who can you trust'.

The next few decades will severely challenge what we have come to expect as fundamental human traits. We will collectively and individually question what it means to be human and as we approach technologies that will make immortality possible we will question not just the meaning of life but also of death.

You can imagine when the technology that today is capable of creating life becomes generally known that this will spark heated debate about humanity playing 'God'. What about those ideologies that believe that God created life; that life is sacred and a divine gift; they won't be comfortable with human creations and even less so with artificially intelligent humanoids.

The above are all very likely. These impacts will result in a general loss of trust in systems, governments, institutions and commercial entities. They will also make people very aware of their individual reasons for being; what gives meaning, 'telos', to their lives.

Diversification of the Species

The most disturbing development however will be in the accelerated diversification of the human species. Today, inequality is already entrenched and a permanent feature of the global tribe. This is a wealth and opportunity access issue and is a direct consequence of our historical 'might is right' and exploitative trade practices. There is no plan to correct this inequality and it's about to grow exponentially.

Access to new technologies will accelerate our human diversification. Initially this will be fuelled by access to beneficial technologies. Technologies that remove diseases, augment human functioning, enhance abilities, extend life and rejuvenate the mind and body will of course be available to those that can pay and are granted access. I applaud all those good people that are well and kindly motivated to help all – but really – there is little in our past that makes me believe that all these benefits will be shared equally.

Such differential access to technologies will be divisive and lead to rebellion in all corners of the globe. This is not my insight; this is well understood by the 'elite' of today. There are serious discussions about a basic income (a stipend, a salary paid to every citizen regardless of employment or wealth that facilitates a dignified life) being held by various governments around the world. This is becoming a popular concept as they begin to understand that unemployment will continue to grow, that millions are redundant and will never do useful work again.

This 'living salary' has been termed 'guillotine insurance' by some of our wealthy elite; those that are growing somewhat concerned about men with pitchforks. They really ought to be concerned because when people are denied access to life saving treatments for themselves or their loved ones, they will be angry; they will most likely take action.

Regardless of any rebellion, protests or even localised revolutions, I believe that the emerging elite and their privileged followers will have the technology, the finance and hence the security to prosper. As we approach true AI, sections of our populations will have the technologies to evolve to meet their future and to enhance their bodies and minds.

These enhanced neo-beings will effectively be capable of indefinite life with unlimited, knowledge, power, awareness and perception. This is not science fiction, it is a logical step in the technology revolution and it is very close at hand. By 2020, this will be largely appreciated and accepted. Such 'intervention evolution' will clearly lead to even more inequality and a divergence of the species. The 'elite', those that will avail themselves of those technologies first will be intellectually superior, resource rich (no change) and the initially dominating sub-species. The emerging sub-species might be categorised as follows:

- The Elite, divided into the biologically, genetically modified and enhanced elite and the AI augmented and biologically enhanced elite (neo-beings)
- AI Robot – classed as sentinel, living beings (not necessarily humanoid but will be ambulatory)
- Semi bio/tech enhanced humans
- The majority in wealthier countries, and
- The remnants of humanity

Those sub-species are perhaps an oversimplification but a good first assessment of what is possible and likely. We can expect these divergent species to further differentiate and develop at varying rates. Equally there is likely to be some convergence with AI machines and AI robots may separate into those with limited 'free will' and those with total free will. It has been commented that once we create true AI that that will be the last thing humans will ever invent.

We might well end up with a large number of sub-species and each of these will have different vulnerability profiles that mean that their ability to survive into the future is indeterminate (existential, natural and artificial evolutionary threats remain and grow). As an example, an electromagnetic storm, some solar radiation event might have a devastating effect on global communications and computing and networking technologies that may destroy the technology enhanced elite whilst leaving those not enhanced with technology unscathed. We can all dream up different doomsday scenarios but the take-away here is that no sub-species is assured survival.

As our environment changes so will the forms and functions of the planet's sentient beings change and adapt. In addition to the already emerging varying vulnerability profiles, an absence of global governance will manifest in wide spread rebellion against the elite. As already stated, the unequal application of technologies will most likely be a massive source of dissatisfaction and add to the elite's vulnerabilities; non greater than the life enhancement and prolongation technologies and I don't think I have to belabour the point that 'why them not us' will be a powerful call to action.

The near future will start to evidence the widening differentiation between socio-economically diverse groupings. The process of differential evolution will accelerate in step with exponential technology and with that, reform of global governance will become increasingly critical. Whilst better, intelligent, technologically aware and moral governance can't stop intentional evolution, it must nevertheless control and manage the expected turbulence which will continue to develop in this, our age of information.

Ubiquitous Information

As we become more and more connected, as the online crowd grows, so will the free exchange of information. The scope and volume of collected data on everything (open, big, live, social networks and hacked data) is already growing exponentially and will kick-up again with the full enabling of the Internet of Things (IoT) when IPv6 is released.

All this data is being retained analysed, sorted and traded and generally becoming available. The free flow of information is bringing about a level of unavoidable transparency. For better or for worse if something can be known it will be. Privacy has been diminished for some time and as more and more people accept and subscribe to a multitude of free applications and software, what little privacy we had is quickly disappearing.

This oversupply of information includes knowledge in general; knowledge once a distinct advantage, now available to all. We need to understand that all things will be laid bare. Like being caught on the spot, actions will be revealed and subjected to immediate fact checking and we can expect revelations for every untruth, misrepresentation and misunderstanding to be immediately exposed. This rapid assessment has never occurred before. It will become another key characteristic of the information age. Instant evaluations, judgements and wide dissemination via the web will strip bare institutions and individuals.

This new openness might well lead to unprecedented levels of collaboration and sharing of ideas, unrestrained by national boundaries, language or location. It is going to become very difficult to 'own' ideas and it is highly likely that the nonsensical patent laws will be phased out. We are entering an era of new empowerment, an era of intelligent use of knowledge and this will demand equality and freedom to be addressed wherever it is perceived not to exist.

Today, we are already seeing numerous past hidden deceptions emerging. As social and political injustices and hypocrisy are revealed, people are reacting with initial disbelief, growing distrust and dissatisfaction. Whilst the effects of openness will initially be endured it will take very little for people to be drawn onto ideological platforms that social media has already shown to be possible and highly effective in order to affect change and provide support to dissenting populations.

The ICT created 'openness', the availability to know it all and to spread and share this knowledge, spells the end of secrecy. This will have widespread impacts on economic practices and will most likely lead to revolutionary changes, changes that will challenge unwarranted profit and advantage gain commerce.

The Loss of Ignorance

Information will eventually overcome the ignorance of the masses. Ignorance is a tool and a precondition of 'force and fraud' politics. By ignorance I mean that human condition that is not aware, doesn't concern itself with the facts or truth on matters that impact their lives and those that simply don't have much knowledge. Those who employ and are well served by this form of advantage politics have most to fear from the information age. Ignorance is the result of comfort and the false perception of permanence. It is maintained by media and failing education systems. I will explain what I mean by failing education below. Media is preserving ignorance because it has mandated an entertaining and commercial role for itself. It generally makes no claim to be educational and it certainly is not. It is directed and controlled by either ideologies or in the case of market economies, by the fraud squad of commercialism.

I am assuming that you accept the general and pervasive ignorance that typifies societies today. If you need proof, simply talk with people on general subjects. From politics to science, basic geography, history to technology you will find that knowledge is scarce. Ask even senior politicians the question of what happens when they send an email? You might be surprised to learn that most of us know very little about most things. It is not that people are unintelligent, simply they lack sufficient interest to know, to learn or to concern themselves with facts about the world they live in.

The disturbing aspect about ignorance is that it starts at home. Whilst it is fashionable to blame education systems it is unfortunately the abrogation of responsibility by parents that is mostly at fault. This is not new; it has been a generational trend. A trend that is self-perpetuating and cumulative – ignorance begets ignorance. Ignorance can be explained in terms of a lack of leadership, a lack of positive influence both individually and socially.

It is necessary to excuse individuals; people have not happily instituted ignorance. We need to acknowledge the causes of ignorance and the existence of inherited indifference. Whilst the abundance of information grows and becomes widely accessible, simply having information does not make for intelligent people or the intelligent use of information; that depends on education and a live-long commitment to grow and learning.

The age of information, the increasing openness and exponential growth of technology we can expect gradual lifting out of ignorance. We can confidently expect to see a social insistence on facts and truths as people develop intolerance to spin and propaganda.

The information fuelled general awareness of ineffective governance will move people towards new ideals of governance and a more egalitarian meritocracy particularly as disruptive effects of technology start to impact widely. Whether it is the likely shortage of employment, the loss of privacy or the lack of hope and direction, people will find a strong voice and social platforms to make compelling and powerful demands on leaders. Just as innovation is returning to first principles (as opposed to building on what is) so people will re-examine the scope, functioning and service of governments. We will certainly see a growing insistence of representational governance where people demand what is to be provided.

The Pursuit of Intelligence

Although much of the world's population lives in ignorance and many will be well and truly left behind in the information era, there are those that are determined to enhance human intelligence. This is partly a response to the acknowledgement that our human brain is a linear, as opposed to an exponential, thinking organ.

In the distant past, our minds could cope. Early man was able to know and share all human knowledge in his social group or tribe. There was little that everyone in the group didn't know or was being taught. As social groupings thrived and grew in number, man specialised and started to focus on specific functions; man diversified and gained specific to role knowledge.

Today, that diversification and degree of specificity means that we as individuals simply can't know what every other human knows. We humans have developed coping mechanisms that allow us to accept as assumed fact and truths all that we cannot learn or verify. This is the heuristic mindset that lets us accept from others what we don't understand and presupposes us to assign a reality to everything in accordance with our beliefs.

225

To overcome the linearity of our mind, or simply the human thinking and retention limits, there are two current approaches being developed. One is the pursuit of strong AI and the other is human brain augmentation.

I've discussed AI in earlier chapters; there is much work and desire to develop a designed and engineered intelligent machine to serve mankind. Whilst true AI appears to still be a decade away, there is growing caution about the perils of creating a smarter that human machine. Elon Musk, Bill Gates and Stephen Hawking are among the thought leaders raising the alarm. There is considerable concern about the logical outcome of created AI machines outgrowing their creators. When such an AI machine is built it will most likely be the last 'invention' of mankind because from that point on, AI machines will not only do the designing and development, they will also decide what to create.

Earlier I mentioned differing vulnerability profiles for the emergent sub-species of humans. Clearly, AI machines will have their unique vulnerabilities and these will be quite different from those of the biological humans. You can imagine the AI machines, being fully cognisant of their needs, having their own purpose and direction might well alter the local or even global environment to suit their needs. AI machines, operating at the speed of light, self-sustaining and having 'out-evolved' their human creators will not ask human beings what they should do, just in the same way we humans do not check with apes or cockroaches (who we out-evolved) on what they would like us to do. True AI will utilise the planet's resources to its own ends and might well become a cosmic explorer, a 'universal cancer' that may prove not to be benign.

So whilst, strong AI poses significant human existential risks, weak AI, still capable of massively outperforming the human mind, will be controllable and taskable by mankind. Weak AI, unlike strong AI, will not have free will and as such will develop into an excellent tool for humanity – perhaps that's as far as we should go with AI?

Education

Perhaps the most vital human activity to lead into the future, education will undergo profound change. It has for several decades now been under relentless attack for perceived failings; what has had much less discussions are the reasons why it is apparently failing.

Since the industrial revolution, education has been pressured to cater to the needs of industry. The early sixties commenced the era of material value, instant noodles, TV dinners and the desire to own more than the neighbours next door. Industry demanded a readymade worker and education was to provide. This started the sacrificing the academic imperatives of education; to enlighten, foster intelligent consideration based on factual knowledge.

In most western nations, social, consumer driven change, soon required both parents to work if possible. No longer being able, and in some cases not wanting, to nurture their own children, the demands for teaching everything, including life skills such as driving, sex education, discipline and even behaviour, was transferred to schools. During that same time 'student rights' became a societal issue and schools lost the ability to discipline their charges.

Not only were educational institutions forced to ready students for work and to prepare the child socially they were also under increasing funding pressure from governments. As demands for this and that curriculum inclusion grew, more subjects deemed necessary, and timetables filled, students were required to choose. It has now reached the stage where some students could opt out of learning say mathematics or science and still graduate!

Little wonder that many institutions are questioning what to teach. Society has abrogated virtually all its responsibilities to schools, demanding first class education whilst indulging in the 'let them be children' ideology that seeks to prevent early learning for some perceived notion that children should be protected from any kind of intellectual pressure. This ridiculous notion that the young should not be tested or subjected to comparative assessment has crippled education.

Arguments were made that pressure on children to perform was detrimental to the child. It was held that they should not be made to compete. Testing, formal examinations of any kind were seen as particularly bad for girls who might be menstruating at that time and obviously couldn't perform. So, education was required to deliver all things provided children were allowed to be children, were not disciplined, tested or comparatively assessed. Not surprising then that real life shocks such as employment interviews, competitive sport and business challenges are finding the young severely unprepared.

Today, the same people who have steadily transferred responsibilities to educators are judging education as failing, declaring that the wrong subjects are being taught (CEOs want innovators, thinkers, problem solvers), that the 'data transfer rate' at universities is to slow, that degrees and higher educational qualifications are worthless and that industry in general is more interested in 'show me what you can do rather than your potential'.

The pressures on educational institutions are great. CEOs demand outcomes, governments resist the costs involved, leaders seek advice, thinkers demand attention to the frontiers of technology and parents demand parenting. All considerable and mostly undeliverable whilst in the west at least, we seriously undervalue our teaching professionals. Our teachers need to be the wisest and most intelligent people of our society and we need to honour respect and reward them because they will shape the people that will shape our world.

Looking at what is to come in education we need to acknowledge and answer a few common arguments. The first is that we don't need to know much – computers and machines will do the crunching and most of the work; really? Early computing taught us a lesson: who interprets computed results? Who has the experience or 'feel' for what is right?

As a military officer, I would ask some of our brightest recent engineering graduates to make me understand a 'newton' (remember, the unit of force).

Whilst all could give me the textbook definition they had no concept of what it actually is or feels like. I told them to stretch their arms out to the side and imagine that they were holding a good sized apple in that hand. I told them that the weight they felt in their hand was a good indication of what a newton feels like and that if they ever again described a load of say 1000 newtons they now had a measure to understand the magnitude of such a force. We do need to understand science and retain a feel for meaning.

Clearly we will also need to design, invent, develop and programme the expectant myriad of computerised machines that will serve us. We will need to retain the ability to certify and maintain such machines and that will require all the knowledge we can muster. There is much hype about additive manufacturing and particularly in the field of construction. There has been far less debate and consideration to who will certify these structures for occupation and statutory compliance. With medical implantable devices will their certification not require knowledgeable human input?

The second common statement about future education is that all knowledge is available on the internet so we don't really have to learn. Well, true; nearly all things are now knowable as never before. Mind you we've had libraries with impressive volumes of written knowledge for a long time but that didn't make everyone capable or intelligent. There is an undeniably impressive immediacy in our ability to know, to find out facts on just about anything. IBMs Watson 'knew' a lot of facts and could win Jeopardy but is still an unintelligent machine. It could not use knowledge in any practical, applied sense.

Information is simply not enough. No matter how much information we have, it is absolutely limited without consideration. In the intelligence world, we were taught the difference between information and intelligence. Anyone can get information but until that information is applied against an objective, is considered and assimilated into the 'neo cortex' (planning, conceptualising, scheming etc.) and networked with all other related knowledge, it is just information. Facts and information is indeed everywhere; intelligent use of it, is however quite rare.

Education is more, much more than the transfer or rendering of information. It is the process that integrates information within the mind, establishes its application to the 'knowledge' base and enhances the abilities of the human mind.

The future of education, just like anything else, is difficult to predict. The merging and converging sciences now so evident in exponential technologies will continue. The current trend of cooperative linkages through the internet, the sharing of ideas and spreading transparency of developmental work will continue and yield spectacular results. We should expect massive outsourcing of online learning to be a big part of education and learning but the traditional educational institutions will also thrive.

We should expect that IT, rapidly becoming a universal pre-requisite skill, will become a base fundamental for all fields of study. The need for academically robust and integrated scientific study will also increase dramatically. As technology enables ever more innovation there will be a critical demand for the expertise to realise and manage that potential. Perhaps the new education systems will be multi-faceted and consist of academic super universities focusing on pure knowledge, online learning that focuses on supplementing valuable and scarce university 'face time' and 'how to' learning.

We are very likely to see the establishment of specialist IT universities, the realm of the IT nano niches as well as catering for the army of needed technicians, coders, programmers and data specialists. It would be logical that these IT universities would include technology hubs as a direct interface with commerce; hubs that would facilitate the free interchange between theoretical possibilities, technology solutions, requested research and practical applications. Further, it is likely that these universities will maintain open crowd sourcing links drawing on the global online community for ideas, solutions and innovation.

Rather than diminishing in importance, I believe that traditional universities will re-emerge stronger than ever. The need for highly skilled and trained professionals will only grow. I believe that today's universities will again focus on pure learning, unsoiled by demands of industry, rather driven by intellectual pursuits. As traditional commercial structures fail through disruption so will the formulaic demands for graduates to be skilled to meet specific industry needs. Because industries will most likely rise and fall on a much shorter cycle, there will be no standard model graduate for institutions to churn out.

Back to Basics

Amid the mounting and often sheepish criticisms of educational institutions there aren't many offering viable alternatives. Knowledge, true and contextualised understanding of everything that defines our objective and abstract realities is mission critical for all who want to matter in the future. Many in our society are already redundant, many are intellectually left behind, many will never contribute to society and a growing number will become more and more reliant on systems and institutions to assist their survival.

As technology, automation and the displacement of people from routine work marches on relentlessly what is the need for education? What should we be advising our offsprings to ensure that they too will not become redundant?

Much analysis is now being done to determine what advice to give to secondary and tertiary students today; advice that will provide them the opportunity to work and to contribute to society in the future.

The report 'Technology at Work v2.0'[60] spells out what reflects an accepted understanding. To be fit for work, to meet the requirements of the next decade, the majority of students need to study computing and IT and STEM subjects (Science, Technology, Engineering and Mathematics). These requirements follow extensive analysis of needs and a thorough understanding of the impacts of automation. Given that the above fields of study are required within the next decade how are we placed? Do we have these students, now in primary school, ready to take on these courses?

Not only do we need students to study the above fields we need them to study these at high levels. More than eighty percent of all future jobs will require higher, post-secondary education with half requiring high tertiary level qualifications and a substantial portion of these needing to be at the master level. Again, are we on target to meet any of these requirements?

Just as there is much analysis of future study there continues to be a growing mountain of advice on the most important and relevant skills needed by the workforce of the future. Now here's the surprise; it's not basket weaving, expressive choreography or finger painting – it's hard thinking ability!

These are the key skills to be developed:

- Sense-making
- Social intelligence
- Novel and adapting thinking
- Cross-cultural competency
- Computational thinking
- New Media Literacy
- Transdisciplinarity
- Design mindset
- Cognitive load management
- Virtual collaboration

Note that they are all intellectual, thinking abilities and associated soft skills.

There are those that believe that 'online' schooling, the ICT enabled access to information, can replace formal, institutionalised learning. Well, it can't - the thinking, the analytical and deep understanding skills that the above list details cannot be arrived at magically. They require dedicated and sustained learning; time for assimilation, explanation and exploration; above all they need enduring human interaction, tutoring and mentoring.

Traditional learning institutions, form primary to post graduate levels, do need to adapt and change is clearly overdue but they must be allowed to teach knowledge. Gone is the algorithmic definition of the student outcome required to fit the industrial age model.

Our educational institutions need to reject the calls of industry that demand a work ready product. These commercial interests have for too long abrogated their duty to take a generally educated person and to intern, to train and develop them to suit their specific requirements. They have through paltry support of educational institutions fed at the trough for long enough. We need to again respect our teachers and educators and allow them to design curricula and teach to impart true knowledge.

Back to basics isn't a slogan. We need our future populations to be highly educated; that requires mastery of so much information. Never has the amount to be understood been so great. We have extended our live expectancy from around 40 years (200 years ago in Europe) to nearly twice that today. During that time we have drastically increased the World's knowledge and hence the material to be learnt and understood.

Have we allowed our young the additional time? No, we've met some of the time pressures by making some core subjects 'electives' and we've created learning environments that simply leave too many behind. The failure of fifty percent of American college students to graduate should be an alarming indicator that something is very wrong. What if we gave students an extra two or three years and allowed educators to design achievable content? I can hear the howls of protest; we cannot afford the time, the cost would escalate etc. My point – we cannot afford to leave so many young and capable people behind – we are failing them.

There is one other point in 'back to basics' that must be stated. It has been a human trait, an evolutionary necessity to have the wisest among us teach our young. We've correctly seen this as our highest duty; to nurture, care for and teach our young is our inescapable debt to them. How is it that the vast majority of the western world has introduced the obscene practice of making the young pay for their own education? This is the most shameful abrogation of our prime responsibility; it says simply that we are so preoccupied with self-interest, that our own comfort and pursuit of selfish goals is more important than taking care of our own creations. That's pathetic!

The Future of Work

There will be an abundance of work but there will be a massive mismatch of skills; the huge number of people soon to be permanently redundant will not be able to do the work required. That's it in a nutshell.

I mentioned earlier that I attended the executive programme at Singularity University. *[Just in case you care, that says nothing about me or my intellect, I was there because I paid to attend – it was a value judgement – a good one!]* We were privileged to be addressed by Peter Diamandis and it was great to be able to ask him questions.

To the question "What's the future of work?" he gave a frank reply:

"I don't know. There are three schools of thought:

- Technology always creates more jobs
- We'll have a near term fall-off, then level out
- We're screwed

I think we're headed to technological socialism…the concept of a career, is a fairly recent social construct."

So that's the answer from one of today's thought leaders and he's right. Nobody can know what the future holds. There are so many possibilities and diverse outcomes but one thing is certain, there will be change.

To those that cite the example of the car replacing horses – get real; the coming technologies are hugely more impactful. Our world has morphed from the process driven industrial paradigm to one that is no longer formulaic. The worker of the future is no longer a cog in the machine; rather he is a product or service to be called upon when and where needed. There will be increasingly less permanent or even long term employment. As Peter Diamandis points out, the concept of a career is indeed a fairly recent social construct.

I'm not going to detail a list of jobs of the future. Such lists are readily available on the internet and are for the most part little more than guesswork based on a failing understanding of the extent and rate of imminent change.

The earlier detailing of 'telos', our defined and adopted purpose becomes very significant now. How many of us can disassociate our purpose from our work, our title, our perceived position and 'standing' in society? What does make a life worth living; what is fulfilment? Most of us think of ourselves as our 'job' title.

Ask anyone to tell you about themselves; they will almost certainly tell you what they do for a living; the 'mere' housewife (or househusband) will almost apologetically state that they are just that. Unfortunately this is how we've developed a sense of worth in our societies. That will change and it will not be a comfortable transition.

The question will no longer be "what do you do for a living" rather it will be "what are you living for".

What then can be said about the future of work? How will anyone add value and be rewarded. Without detailing lists some future spheres of work are obvious. Pick the three most aggressive technologies (AI/robotics, genetics and nanotechnology) and their driver (computing) and apply them to every human endeavour and you'll understand the future of work.

The current thirst for data and its commercialisation is another huge clue. There will be significant employment for IT and data scientists for the next few decades – until at least the mid-2040s when we might achieve true human level AI.

There will be growing demands on people mentoring and psychological skills. As change impacts ever more dramatically there will be a wide range of work in assisting people to cope. So from industrial psychology, to communicating change and alternative ideologies to counselling the disoriented there will be much to do in those areas.

The other rapidly growing job market will be in security. The whole spectrum from testosterone rich security guards, police, Special Forces, counter terrorist practitioners to intelligence agents and cyber policing this will be an area of high employment into the distant future.

I started this section by stating that there will be a severe mismatch between available work, which will be plentiful, and those seeking work. This is no Nostradamus prediction. Already we have massive unemployment and equally huge lists of vacancies; unfortunately the necessary skills just don't match. We need to acknowledge that retraining is not the solution many would have us believe. Do we really thing that a fifty or sixty year old manual worker can attend a government sponsored course and become an ICT professional in short order? No. we need to accept that unless one's aim is to be at the lower end of the service industry then education; robust education, is the only sure path to remain relevant in the very near future.

To be worth employing, individuals need to be able to add measurable value. Value that is based on real knowledge and deep understanding of both the objective and ever more complex artificial world we are creating.

Uneven Change

As we settle into the era of information and witness the disruptive adoption of exponential technologies our differences will become greater. The remnants of the industrial age will linger for decades reflecting the varied states of development around the world. Whilst some cultures will skip entire technological paradigms, most will be left with a mix of the old and the new.

I've pointed out previously that we 7.4 billion humans are not a homogeneous group, nor are our societies or nations. So, whilst there is much talk of change, and indeed change is coming, it will be unevenly distributed and experienced. Access to the most beneficial technologies will naturally flow to those that can afford it.

The most disruptive elements of the changes to come are those described above; the uncertainty and awareness that life is irreversibly being changed and the realisation that these changes are being introduced without any global governance. It is clear that we are developing technologies that are massively disruptive and that we are doing this without a plan, no control and no brakes.

There will be large communities of people that will believe themselves to be 'safe' from disruption. Their linear minds cannot, or don't want to, perceive the implications of true AI and the enabled transcendence of human limitations. Their certainty and comfort will however be short-lived.

The key message here is that change will affect us at different rates. Not all manual work will be instantly automated. Just because robots can do something doesn't mean they will. Automation will happen but again it will happen at different rates. The impact of governance will also become a key issue. Nations understanding the disruptive impacts on employment may well, in their national interest, legislate to limit automation altogether. Such nations may simply not have the resources to pay a living salary to its citizens and may therefore decide to restrict technology to preserve employment.

The most divisive and socially crippling aspect of the changes to come will be this appreciation of the growing inequality; an inequality once based on wealth. This 'old' inequality was somewhat tolerated; the concept of the privileged few has after all been around for millennia. The new inequality of access to life changing technology will however not be so readily accepted.

We are only a few years away from 'designer babies' and altering the physiology of adults. When the general public becomes aware of technology's ability to wind back the aging process, to rejuvenate the bodies and minds of old people to their prime, they will insist on access.

Change will be experienced at varying degrees and rates. This will make our already diverse realities even more different and rapidly introduce new scales of inequality. So whilst we may not all personally experience much immediate change it is nevertheless happening and to most, when it manifests, it will appear as a tsunami. The warning signs however are clear and obvious.

Organisational Change

When I initially described exponential technology I detailed the 6Ds and in particular 'disruption'. There are numerous examples where this disruption is causing havoc amid established industry players (see Chapter 6).

It is not surprising that eighty percent of Fortune 500 C-level executives believe that their companies will be disrupted within the very near future; the remaining twenty percent accept that they will be within 5 years. Nor is it surprising that business analysts have speculated that forty percent of these Fortune 500 companies will not be around by 2020. They are all acknowledging that change, big change, is about to totally transform business and whilst many are scurrying back to the change table with the view of disrupting themselves before someone else does – none are certain of their ability to survive.

Just as our linear minds will find it increasingly difficult to deal with exponential technologies and the world that is being created so too will linear organisations. By definition, linear organisations are typically capital intensive with considerable physical infrastructure and facilities. Scaling up production and product or service development takes a long time, involves risk and basically seeks to anticipate the client's needs and then selling the product or service to them. Such organisations are very hierarchical, employ significant numbers of people and value the knowledge worker. Their very structure and organisation makes them slow to change or adapt; they are rife for disruption.

Today's large organisations reflect industrial revolution design; large, compartmentalised, process driven, resource heavy and with compliant knowledge workers and management they have enjoyed the world they once dominated. All this is of course coming to an end. Size and its inherent slow reaction to change are being usurped by the new much more agile and tech-savvy organisation. Today, anyone, sitting wherever in the world, with a small amount of money, even a credit card, can establish a multi-million dollar enterprise. Playing in the big game, the game with real scale and global reach has been enabled by technology and our connectedness.

In the past there have been numerous instances where large corporations stepped out of their own organisations in order to achieve significant results. One such example is Lockheed, who in response to the US government's request to build a jet aircraft in 1943, setup a separate operation which isolated a team of specialists to work on that project with total freedom. Less than 150 days later a jet fighter was delivered; an astounding achievement and a testament to goal setting and the application of innovative pressure. The concept of big objectives, isolation and rapid iteration of ideas and prototypes was again used much later by Apple. In the early 80's, Steve Jobs used that same principle in the development of the first Macintosh computer.

Both these examples should have been a wake-up call to industry in general. How is it that to achieve time and innovation sensitive results these already high performing organisations turned to 'skunk works' [4] to get the job done? Surely it would have been a correct conclusion that the large, highly structured and controlled organisations were not ideally setup to innovate.

Anyone can access the world's best minds, the most competent coders, algorithm designers, programmers, application developers, scientific minds, engineers etc. and harness their abilities for often minimal cost. Most importantly, you can gage feedback and consumer sentiment immediately. No longer does anyone have to spend years, millions of dollars and considerable effort developing or refining a product in the hope of getting the consumer interested. That can all be done with little more than a concept. Crowd source the idea, the concept or the initial design. Let the crowd refine its specifications; they will also buy-in and your market will be fundamentally established before even the proto type has been built. That is what the large, linear corporations are facing; little wonder they are worried about their future.

Those that understand how to take advantage of the online crowd, the impossible is not only possible – success is almost guaranteed. I highly recommend two books to you. Firstly, to get a thorough understanding and an experience based 'how to' achieve in the new information age you must read 'Bold'[61]. This will explain everything; from crowd sourcing through to crowd funding and it will give you all the contact details, web sites etc. to check it all out for yourself.

I will go on to detail what I believe new organisations will look like and how they will operate but you really should read Salim Ismail's book 'Exponential Organisations' [3]. This is a definitive work; it demonstrates who and what are changing, what organisations are doing to become exponential and it even details basic job descriptions for C-level executives. It is a must know book if you are serious about understanding the organisational drivers for the future.

Perhaps the most significant changes we are likely to experience in the very near future is in the way organisations are structured and how the commercial world will change. I draw a distinction between commercial and government organisations because as one is forced to change, the other, government and statutory institutions that must implement policy and enforce legislation will change to a far lesser extent. We will see a marked difference and a divergence of not just organisational structures but also in key operational practices.

Permanent employment will be largely done away with. Only those in military, policing and government service are likely to experience any form of permanence. This will establish a divergence between those virtually guaranteed work and those that are basically the 'as required' service or product providers. This divergence will have far reaching consequences and will initially see fierce competition in filling the shrinking public service positions until society develops to the extent that all will receive a living salary, working or not. This is being considered already and as unemployment grows and people now made redundant are increasingly realising that they will never work again; such new solutions are becoming pressing.

A living salary for everyone won't of course satisfy all, nor would it provide for societies needs for long. It begs the question: if we all get a salary for not working then who cleans the toilets? Incentives to work will be required and that must offer some worthwhile gain; gain that is not available to those that do not or cannot work. So whilst we will have the safety of a living salary, society will continue to provide rewards for those that do work and further divergence shape our societies. Fundamentally then we have three basic groups; the long term employed public servant, the not working living salary recipient and those that do work. The group of 'workers' will consist of professionals, those skilled in trades, the providers of specialist services, academics, teachers and scientists. This group will remain a high value (and be valued) to society and have long term employment.

For those that do work, the nature of how work is done will be very different and to understand that, we need to consider just how new organisations, commercial entities will be structured and organised.

The golden rule of 'he who has the gold makes the rules' won't change. Those that espouse a totally flat organisational structure and advocate a 'no management' model are simply wrong. The gold however will be in two forms; those that actually have investment capital and those that have the ideas to generate large incomes. There is no doubt that capital will find talent and there is even less doubt that ideas held by those that can make the 'virtual real' will attract massive capital. As traditional blue chip companies gradually diminish there will be huge surpluses of capital looking to be allied to innovations. The entrepreneurs of the future will be sought after and be the universal creators of wealth.

The person with the funds, the idea and the determination to provide a service or product will implement an organisation to achieve their objectives. A small leadership team is likely to be the only permanent human resource employed. Everything else that is required to develop, market, service, manufacture, distribute etc. will be engaged on an as required basis. For complex operations, we can expect sub groups of largely self-organising teams to emerge, to do their job, to interact across all levels both within and outside the organisation, and to dissolve when their tasks are completed.

To visualise this model imagine the surface of a body of water. Now imagine a few scattered but heavy drops of rain hitting the surface. Note the ripples forming and spreading in concentric radiating circles – these are influences and actions. Imagine more drops falling within those concentric circles – these are the sub-groups of 'employed' resources. They too cause ripples and interact; they also have drops falling into their concentric circles of influence. By now you can imagine the surface as being quite dynamic, lots of drops impacting, dissipating and fading in a continuous and relentlessly changing but predictable fashion. That is what the modern commercial world will look like. Structures, organisations will form to cooperate, to share skills, to innovate and produce.

The above will be the structure and operational behaviour of the majority of organisations; these are the exponential organisations of the future. They may be long term; where the core remains largely permanent and objective driven sub-groups may equally be effectively permanent. Start-ups are already exhibiting these traits – structure and organise to achieve particular outcomes then change and move on.

Some large enterprises will survive and be enablers for the more short term objective focused companies. Companies such as Ali Baba, eBay, Amazon will continue to hold dominant positions. They are to all intents and purposes already exponential and agile and are providing services that are not easily disrupted. Equally, we can expect distribution companies and companies that provide start-up services to strengthen. Companies that provide IT coding, software development and those that can access the market through deep data analysis and customer identification will prosper.

A key feature of future organisations will be the interactions with the crowd, the online crowd of between 5 to 7 billion people by 2020. Ideas will be shopped to the crowd. These are pre-commercial ideas seeking to determine the appetite or need for a product, service or application. Exponential companies will seek to harness the power of the crowd to source innovation, design and to solve complex problems. By engaging the crowd early and continuously, entrepreneurial organisations will rapidly cycle through ideas, develop and test prototypes and be prepared to test and to fail fast and often. Engaging the crowd will also prequalify the crowd. Not only will they clearly understand the crowd interest of any particular product or service, they will actually have pre-qualified customers.

But crowds are more than consumers. The online crowd will be the primary source for innovation and problem solving. Try it even now. Post a problem or question and see the response. You will find that there are thousands of people in any field of endeavour willing to contribute their knowledge and time to deal with your issue – and most will do it for free. These are the intellectually capable and often frustrated people who have so much more to offer than is being asked of them; they are the world's huge but untapped cognitive reserve and they will become an integral part of future innovation and commercial processes.

These changes are all predictable and are manifesting now. The challenge for government is to firstly recognise these trends and to prepare and enable a smooth transition for the changing commercial world. With only minor exceptions, certainly in the west, governments appear to be asleep at the wheel. In countries like Australia, the government is consumed with minutia; they spend inordinate amounts of time dealing with simple to resolve matters without any acknowledged awareness of future challenges and even less preparedness to deal with it. This will in time be recognised and recalcitrant, unaware and unprepared governments will be severely tested by an ever more impatient population.

By far the greatest challenge however is to the individual. The need to define what is of personal value, how to live a life that may never offer employment, how to participate and how to prosper will be the preoccupation of the majority of humanity. Consider the current inequalities and the emerging hugely beneficial technologies that will serve those that have the means of access and you can imagine the stresses that that will induce in those that are left out. In following chapters I deal with the need for self-direction and leadership. These will become the new information age survival tools. To determine your purpose, to define your influence and direct yourself are the paradigms of the future.

In summing up the near future it should by now be clear that we are in for an exciting time. There will be significant transitional confusion and pain. Governments are likely to be very slow in acknowledging permanent unemployment and the need to provide living salaries for all. As society redefines wealth and values and individuals come to terms with their motives, values and incentives they will do so in an environment where the benefits of exponential technologies will become ever more apparent and desirable. Just as apparent will be the realisation that without global governance, any equitable access to such desirable benefits will be denied to the majority.

We should really be prepared for considerable political instability. Where governments are slow or incapable of addressing emerging issues, particularly equality of opportunity and access to technology, they are likely to meet considerable social unrest and even violent revolution. The future will be one of mass disruptions. They will fundamentally be human evolutionary disruptions but the threat of natural and environmental disruption such as the impacts of climate change, will remain ever present.

Is the likely future to be one of 'disruption'? Ask yourself what will change; alternatively what do you believe will not change? How are you placed to deal with it? What are your options? Do you accept that we are all going to be tested or at least significantly impacted by exponential technologies? If you are personally wealthy and comfortable and believe that you are immune to change or that you might simply isolate yourself from any disruption – really, are you certain, your family and the people you care about – all OK?

The near future is one dominated and disrupted by information and exponential technologies. We will see the emergence of new values and a newfound prosperity of time. This era of massive transformation has begun. We need to change individually, socially and globally if we are to establish a common ethical and moral world. We have no choice.

CHAPTER 11

Self-direction

"If you don't design your own life plan, chances are you'll fall into someone else's plan; and guess what they have planned for you? Not much. (Jim Rohn)

What is Leadership?

Let's define leadership. The world renowned author, speaker and leadership expert 'Dr. John C Maxwell' explains that "leadership is influence; nothing more, nothing less". I can't fault this definition because I accept that this is exactly right.

Isn't that what every leader does? Don't leaders influence the behaviour and actions of those they lead? Observe any group of people and note how they agree on anything. There are usually one or more people in the group that will suggest an action or activity that the group accepts and implements. How does this happen? It happens because they were influenced and therefore led to act.

Simply, leadership is that quality that manifests itself in the ability to influence outcomes, initiate actions and drive behaviour. And here's the most important take-away: we are all leaders! Anyone who interacts with someone else exerts an influence; even passive acquiescence is an influence, eating a meal that someone has prepared is an influence. A husband saying to his wife "you led me to believe that you liked housework" (silly example, I know) was influenced to come up with that conclusion. There was no leadership contest or appointment, the husband was nevertheless influenced and led!

In coming chapters we'll look at leadership fundamentals and I hope to show that these leadership qualities apply to us all. Leaders aren't a different breed; they are simply more focused on applied influence on group, social or organisational objectives. From this point on, if you are not in a formal leadership position, I would like you to identify as a leader. The global issues that exponential technologies are enabling concern all of us equally; however, acknowledged leaders, because they are in a position to influence a wider domain, are singled out just because of that fact.

Leadership Starts with You

What if you accept that you are a leader but say "I have no one to lead"? Well I say you do. You need to lead yourself first. Leadership does actually start with you.

'Lead yourself' sounds a bit odd doesn't it. But, consider the developmental journey of the average child.

The child's early years are basic, formative stages focused on physical growth and developing mental awareness. The child is intensively managed, nurtured and led to a stage of development that renders it fit for formal schooling.

As soon as the child commences institutional learning, the child is guided within agreed limits as to what it is to focus on, what it is to acknowledge as real and important and what it is to ignore. Schooling provides a roadmap, a curriculum, boundaries defining what to do, what performance requirements the child is to accomplish; it in fact leads the child to act and develop in some prescribed manner.

Formal education, might be seen as leadership applied to a child in order to meet a set standard of personal development. It can be argued that as the child becomes a teenager this external leadership starts to be challenged to varying degrees. This can give rise to a form of rebellion which questions 'why do I need to study this or that'; why should I keep doing this...'

I'm sure you can think of many examples where rebellious attitudes have been displayed, when young people start to question the externally applied requirements to act in a certain way, to conform or to otherwise 'do as they're told'. In other words they are experiencing and rejecting (at various levels) being led to act in ways that they have not necessarily agreed with.

This is of course a totally natural and vital part of developing into an adult. We need our young to take personal ownership of themselves. As the young become independent they want to be rid of institutional constraints and external direction. They are keen, entitled and have the right to self-determination – they are ready to lead.

The young adult now starts to evaluate the world and their place in it. The fact that many become confused, bewildered, intimidated and sometimes lost is a sad reflection of what they perceive. Little mentorship, a lack of preparation for self-determination coupled with the sudden freedom of action presents no clear direction to the young adult not prepared to lead themselves through this stage of life.

So why is 'leading yourself' important now?

It's important because the world has changed so much over the last half century. Let's acknowledge how things were fifty years ago and look at how this compares to today.

Then, work was easy to get. If you went to university you were practically guaranteed a well-paid job and that job could be held for decades if not for life. Similarly if you become skilled in a trade or specific service related profession, your skills were appreciated and loyalty was given and expected. You could count on reliable work; you could count on a steady job, you could expect promotion or expansion with experience and enjoy the benefits that that brought. Your career path was not only predictable but also fairly achievable.

Now, it has become almost impossible to get meaningful employment without experience or an appropriate connection. If you do gain employment it will most likely not be permanent or full-time and there is no guarantee that you will be kept employed for any length of time. You may be viewed as a task or objective based resource that will be released on completion of that task or objective. You have become a defined value resource to be paid and kept on only whilst your resource is required. You are no longer a long term asset to your employer- you, your abilities will be utilised only when needed. This is why you need to develop yourself as an asset, someone that can add value to others. What will you be able to bring to the table? How will you be of service?

Then, there was time to consider, to consult, to research, to talk, to read, to relax, to enjoy quiet time, to be uncontactable – simply you had time to live. If you wrote to a relative or friend on a different continent it could take two months for them to get your letter if you couldn't afford air mail. Telephones were usually confined to the boss' office and reserved for only very important communication. If you needed information it took considerable time to get from books in the library or someone who knew - as long as you had access, made and spent time to talk and learn from them. Knowledge and skills were valued, it made people valuable and it gave them a sense of self-worth, accomplishment and respect. The individual could actually rise in social status and be rewarded for what they knew and could therefore do.

Now, the pace of life has increased so dramatically that unless you make time you have no time to spare. Most people living in the modern world have insufficient time to do or consider anything in appropriate detail. Decisions are now made in haste; electronic texting, messaging and emails are direct and in most cases require immediate responses. Thinking time has been reduced and whilst communication is very quick it is not subject to the deliberate consideration and evaluation it might actually warrant. I'm not saying that this is in any way bad; simply that time is scarce and people are forced to act quickly.

Knowledge or subject matter expertise about most things is no longer the significant differentiator that added value to a person because knowledge is only a 'Google' away. Being knowledgeable is no longer as valuable as it once was because there are numerous, almost instant, alternatives to gain such knowledge. This has resulted in a perceived lack of value of subject specific learning. Having been educated in say the three traditional subjects of primary language, mathematics and science is not that special when any desired fact can be quickly discovered, any text can be translated at the touch of a button or a voiced instruction to a machine. So subject knowledge is simply not as precious as it once was.

Education enabled the educated to contribute in very direct and mutually beneficial ways to society; today, particularly in well developed nations, education is often seen as missing the mark, of not preparing the young for 'real life'. Thankfully, there are a number of progressive countries (Finland for example) currently experimenting and in some cases implementing updated systems. Educational institutions that stubbornly hold on to their curriculums, content and out-dated methods are in fact causing their self-devaluation.

In the past, governance, the conduct of governing nations, whilst never really serving their constituents very well, was at least considered appropriate. Governance today is failing spectacularly. In most nations there is a growing realisation that laws of governance are continuing to fall behind and are reactionary in nature. That those laws that govern the internet, media, freedoms of speech or association, welfare, taxation and so on simply cannot address the complexities of today's very dynamic world.

One could argue that the inability to proactively manage any large corporation or nation is a direct result of the fast changing world we live in. Clearly it takes considerable effort, process and time to position a corporate or political leader. Often, credibility of a potential leader has resulted from some level of expertise or accomplishment. When past expertise, rather than an ability to lead, is the reason for the appointment, little wonder that that individual doesn't correctly anticipate and manage a changing environment.

How does any of the above make a case for self-leadership? Well, I ask that you acknowledge that the old, traditional environment although constraining and limiting, provided a roadmap for life. This roadmap however hasn't been updated to take into account the already much altered environment and it certainly does not provide meaningful guidance for a rapidly changing and dynamic future. Today, life options, pathways and objectives that once enabled a degree of certainty about what sort of life one might expect has been replaced by what?

Surly, we all need to understand the situation we find ourselves in. We need to acknowledge the world we live in for what it actually is, not what it was or even how we would like it to be. Base any analysis on reality, anticipate the future and think on how you want to live. What future might you want to live in? Determine your path, your future and lead yourself in building it.

The Default Position

If you doubt the need to lead yourself let's consider the default position. Let's assume that you've not realised that you need to direct, manage, indeed accept ownership of your own life. Say, you were born 40 to 70 years ago and didn't consciously perceive the changing environment; that, like most of your peers you simply 'got on with it'. How might that have worked out for you?

Over the course of the last twenty-five years I've spoken to and interviewed hundreds of people. Whilst I probed the subjects of life, career, family, wealth and purpose, I tried to gauge their degree of happiness, their sense of achievement and satisfaction. I searched for any acknowledgement of regrets about choices and decisions made in their past.

I want to stress that the people I interviewed were mostly professional, well-educated and apparently successful men. Why mostly men? Beside the fact that I operated and socialised in very much male dominated environments I found it easier to probe the thoughts and opinions of men (without the risk of being misunderstood) and I was becoming more aware of mental health issues of men. These men were colleagues working in Australia, Asia, Europe and Russia, in a range of industries in both the public and private sector.

I would ask:

- Who are you?
- Married, children, mortgage; how's that working out for you?
- Are you having a crisis?
- Should you be having one?
- Are you questioning who, why, what you are; the meaning of life?
- Have you decided or concluded what is worth doing?
- Have you thought about what you do or can do to make you feel good about yourself?
- Are you a passenger on a train of bad decisions to destinations unknown?
- Have you taken stock of your situation and have you pondered what, if anything, it all means?

The most telling question would often be simply "who are you?" To this startlingly simple question most answered with a profession in their reply. I'm just a solicitor, a painter, a lecturer etc. Most identified themselves as a job title not as a person. That's sad. It's pitiful because these people have become their profession; life outside work has become a fill-in, a duty and something to be endured when away from work; it is work that gives them status and a reason for living. Most had never asked why.

Questioning further to test understanding of their purpose, what it matters in the long term, what the mission is and where the end benefit of living life as a profession is, again revealed little. The most common responses related the need to make a living and to support or contribute to their families. Fair enough, but how much of 'you' has been shelved to meet, what you think, are your obligations?

One of my standard and evidently most difficult questions to answer has been "what do you do that's just for you?" In almost all cases, when I've asked this of middle aged men, my query was met with silence and a shrug of shoulders. Digging deeper and asking "what do you do for fun" equally met with little response.

Countless, seemingly normal, men assert that they do virtually nothing to just please themselves; rather, they believe that they have become a provider to others and conformists to routines, never consciously considered or accepted. They do because they are expected to do; they are habitually going through motions without challenging why.

What I found particularly sad about the men who admitted to having little purpose and doing nothing 'just for themselves' is that they felt that it was all too late now anyway. What a waste! I had to point out that they needed to enjoy life now, that life was in all probability a singular event, a onetime gift and that there's little chance that they would get another stab at it. I challenged them to face their disappointments and to change their circumstances.

I would ask "so what would you change or undo if you could?" Surprisingly, most spoke about relationship related regrets first before addressing professional choices, careers or business decisions. Many voiced regret that they followed their mind rather than their heart. They had subordinated what they wanted to do or to be. They were influenced to accept and do the right and responsible thing. This is now often described as not following one's passion. Living without 'telos'.

As we live, we all accumulate considerable psychological baggage. This baggage includes beliefs about us, our habits, notions of duties and responsibilities, likes and dislikes; beliefs that impact on our perceived options and abilities. I liken these beliefs to untested and limiting assumptions that we carry with us regardless of any real benefit. It's like going through life in a large truck; the more beliefs we gather the bigger the truck. Manoeuvring an unwieldy truck through life is difficult; it limits you to obvious and direct routes and your ability to go where you might want is severely restricted.

Examining what you hold to be true about yourself may lead you to drop some of that baggage. Don't accept that you're not good at this that or whatever. Understand that you can do whatever you set your mind to do; if one man has been able to do something – so can you if you really try.

Question your core beliefs; accept them as a previously held belief but not necessarily as a truth about yourself. Get rid of any baggage associated with guilt. If you've been wronged, abused or belittled in any way – get over it and move on.

So, most men that I interviewed had little joy in life; they were a job title, felt unappreciated, lacked validation, were still looking for approval and were now bound to doing the right thing. Most had families and had adopted a swag of limiting "must do's and don'ts".

My observations demonstrate what I have now accepted as a reality of our artificial social constructs. The much revered family unit, once a survival necessity, no longer serves us well. Traditional relationships and the sanctity of marriage are becoming irrelevant (perhaps the increasing rate of divorce is an indicator). Too many people are unhappy. Most have never made a conscious decision about the issues they now regret the most.

The above shows what experience many people have when leaving life's decisions, directions and actions to default. Doing the right thing, the logical thing, the expected next step is not wrong but I say most failures are the result of always doing what's right.

Well, doing the right thing is by definition the right thing to do; but who says what's right for you?

What is influencing you?

The default position, the strategy where you simply 'go with the flow' could well lead you to accept a life not deliberately desired. If you don't consciously examine your heart and your mind to set goals and define your intentions then what is leading and therefore influencing you?

Clearly your upbringing, your beliefs, your social and geographical environment will have armed you with values and opinions. You will have adopted core beliefs and consider these to be facts about the reality of the world you live in. Undoubtedly, many of these core beliefs will be factual; some may not be.

You could think of yourself as a data holder. All that you know all that you believe to be true and all that you can deduct from this knowledge that you hold is really - all you know. As you continue to live and learn through experience you may well amend the data in your mind to reflect new information; you will be adding to the data in your mind and most likely, changing your perceptions of the world. Given that this is so, how important is it to manage what knowledge, information and experience you are adding to your data base?

Acknowledging that what you believe you know will influence how you feel, your emotions, your thoughts and perceptions, your behaviour and ultimately your actions, isn't it very important to make sure that the data you accept is accurate?

Media, in all its forms, is regularly the most direct and powerful influencer in modern society. That would be OK if media was presenting facts and made clear distinctions between fact, assumptions, interpretations and opinion. We are all aware of the power of television and most of us enter a trance like state when we are captivated by a programme that has drawn our attention. Unfortunately, television has become mere entertainment and that includes news and current affair programmes. One of the wealthiest media owners in the world once defensively stated that it is not a media responsibility to educate the population. Whilst this may be so, for too many people, media, specifically television, has become their only source of learning about what is going on. Not surprisingly, too many then are educating themselves on prime time entertainment programmes.

'If it bleeds it leads' has long been the catch cry for news services. We have all been exposed to sensational coverage and headlines of the accidents, the terrorist activities, the disasters that occur with reliable frequency.

Regardless of relevance, geographical location or news value, these are the stories that dominate media. If that is what you are exposed to on a continual basis, what data are you receiving and how is that shaping your perceptions of the world?

To make matters worse, the people that are presenting these stories are more and more giving their own opinions on the matters that they are meant to be simply reporting. Is this deliberate to ensure that you understand the implications of the latest headlining event? Is this based on an assumption that presented with the facts you are incapable of your own assessments?

Accepting the 'influence reality', acknowledging where we get information, its quality, purpose and the effect it has on our perceptions and understanding, is critical. We must be conscious of what and who is influencing us and how we assimilate, reject or accept such influence. We need to evaluate our mind resident data and our deductions based on that against our values and purpose, if it is to be of any benefit to us.

The people around you

In this, our present information age, many boast about having hundreds, even thousands, of friends on social media. As live and virtual online networks rage everywhere, more and more people are proclaiming their impressive numbers of 'friends'. Don't be threatened or taken in by this if you don't have hundreds of friends. It's hype and simply misguided to consider such contacts or 'friends' on Facebook real friends.

They are nothing but short-term links, contacts that share the aspiration of popularity, share a momentary common thought or purpose - but they are not friends. The number of friends that we can actually have is limited by our mind. The size of our brain, our neocortex, imposes a cognitive limit on the number of stable personal relationships, friendships, which we can maintain. It's that simple.

The evolutionary anthropologist, Robin Dunbar, established a link between average brain size and the size of average social groups. He found that human beings could effectively only maintain 150 stable relationships. This has become known as 'Dunbar's Number'. His theory contends that we can only manage relationships with about 150 people at any one time and defines such relationships as staying in contact at least once a year and being aware of how friends relate to others. This was initially thought to apply only to the 'real', the non-online world but in 2010 Dunbar showed that the rule applies equally to our online relationships. So, don't believe the Facebook hype; you only have a maximum of 150 friends.

150 friends – really? 150 people that you contact at least once a year. Unless the definition of friendship has been totally re-written I doubt that there are many people who have 150 friends. The average number of friends we actually have is three. If the results of the study 'Social Isolation in America: Changes in Core Discussion Networks over Two Decades' (published in June 2006, American Sociological Review) are accurate then there is the disturbing trend that Americans are suffering a loss in quality and quantity of close relationships. Over the past three decades, the average number of close confidants has dropped from four to two.

So realistically, we don't have thousands of friends; Facebook or otherwise; we can only really sustain up to a maximum of 150 close friendships and in all reality we are blessed if we actually have three friends. How important are these friends in any case. How do they and the people we regularly associate with actually influence us?

"You are the average of the five people you spend most of your time with." – Jim Rohn

There is no doubt that the people that you regularly associate with are a great influence on you. You will accept a 'group think' compromising attitude, develop and accept shared values and opinions, develop your self-image and adopt the behaviour traits and general characteristics of the group. That's all good if your group reflects your intentions, goals and desires; it's highly destructive if that's not so.

Given you will become the 'average of five' it's worth having a close look at your regular associates. Do you like their behaviour? Are they thoughtful, deliberate and responsible? Do they exhibit the characteristics you respect? Are their goals, ambitions and motivations similar to your own? Do you see in them the qualities you desire in yourself? Do they exemplify the person you want to be?

Taking a cold hard look at the people you spend the majority of time with can be a sobering exercise. Do it. Ask yourself what that says about you; what it says about your choices and the opportunities you are giving or denying yourself. Adopting a group identity, associating with people and adapting to conforming ideologies, actions and intentions will influence you greatly. We do all need to be aware of the influence that the people we spend our time with have on us. If you find that these close associations are not reflective of who you want to be – then change it; choose new associations.

In the paragraphs that follow I will introduce the concept of your higher value; your passion and desires. The people you are in regular contact with need to allow you to live your purpose and to enable you to achieve your vision for yourself. The people you spend your time with are not just influencing your thoughts, perceptions and behaviour; they are shaping your future.

Living your Telos

I described our higher purpose, our 'telos', in Chapter 3. We all have one, consciously adopted and aware or not. It stands to reason that knowing one's goals and ambitions is a prerequisite to achieving them; too many of us are however cruising in default mode. It is sadly very common to hear admissions and regrets anywhere where old people are gathered. We can learn much about their laments.

Bronnie Ware[62], in her book "The top five regrets of the dying" lists them as follows:

- I wish I'd had the courage to live a life true to myself, not the life others expected of me.
- I wish I didn't work so hard

- I wish I'd had the courage to express my feelings.
- I wish I had stayed in touch with my friends
- I wish I had let myself be happier

Note the poignant first regret; define your telos and be true to yourself.

Unfortunately, the above regrets are all too common; they are typical of the generations that arose during and since the industrial age. The majority of people accept their particular perceived environment, the circumstances that define their life and never question why; they seek above all to conform to expectations, to fit in and to lead the life expected. How sad then when we question the dying about the why; the purpose and we get answers as above.

It's easy to go with the flow; just do what is expected, let the years pass and never take stock; put off to tomorrow any self-examination of why you exist, what your mission or purpose is. This is what most of us do. I am personally disheartened when I listen to the regrets of the elderly; they appear so pathetic when they admit to having had no good reason for doing what they did; they say they were mostly doing what they thought others expected. What is even sadder is that they had never actually asked these 'others' what was expected. Not even the sense of having done the 'right thing'' then provides any comfort.

A life without telos is a life drifting in the current of events. Bouncing from one momentary influence to another is what a life without purpose becomes.

We could all spend our lives influenced by friends, family, colleagues and even 'on screen' personalities. We could allow these influencers to impact and direct our thoughts, feelings, behaviour and actions; we could just accept these default influences and allow ourselves to be subordinated to the ideologies and purposes of others. We could lead directionless lives devoid of fulfilling purpose or passion. On the other hand, we could take ownership of our life. Appreciating our uniqueness, our divine right to exist and our gift to experience life, we could determine and live our 'telos'.

Organisational Telos

Just as a personal higher value system is going to be vital in defining the marketable individual and future leader, corporations will define themselves by adopting and promulgating their aspirational higher purpose. This is already happening and will become standard practice for exponential organisations.

The modern, hyper connected world allows consumers to 'shop' globally. No longer forced to make decisions based on regional proximity or localised monopolies, they are free to source products and services from all corners of the globe and with that, a new discriminating factor is emerging. Customers can now impose their values, values that reflect their telos, onto the products, the manufacturers and suppliers they choose. That means that decisions will be based on organic production, environmentally sustainable processes, social values, equality, and treatment of employees. Decisions to do business could even depend on whether the organisation pays its taxes or not and is perceived as a good corporate citizen. Individual values will be used to assess with whom business is conducted in much the same way as it determines personal relationships.

It is not surprising that organisations are deliberating on their telos. Described as 'Massively Transformative Purpose' (MTP) by Peter Diamandis (*Abundance 360*), organisations are defining their telos, their mission, and melding them into their profiles. Examples include Google's 'organise the world's information', TED's 'ideas worth spreading', Baidu's 'relentless for ultimate intelligence' and Baidu's Institute of Deep Learning MTP: 'innovation for making a better world'.

How we act, are perceived and judged as individuals or organisations is becoming as important as the product or service being provided. People will buy into your telos or your organisation's MTP before they buy your product or engage your services. Self-direction that personifies and manifests your higher purpose is becoming a 'should do' for individuals. Not directing yourself isn't going to harm anyone but yourself; subordinating yourself to others might actually be of benefit to them; the fact that your life will have been less fulfilling is not likely to keep others awake at night.

For organisations however it's a 'must do'. If people don't accept a company's MTP, if they find it disingenuous, misleading or meaningless, if they perceive the company's conduct to be anti-social or environmentally damaging – they will simply not deal with it. The branding of a company, a catchy slogan or annoying jingle might once have been enough to attract customers; not anymore. People are looking and expecting more and the information age is enabling value based decisions.

I urge you to take the time to understand what you are all about. What makes sense to you, what makes you enthusiastically bounce out of bed every morning and then to follow your telos. Understand that there is no right or wrong answer; you telos needs to be aspirational and sincere, heartfelt and mindful; it doesn't have to be to make the world a better place, to feed the hungry or walk on water. It could be as simple as being a good father, to make people laugh, to excel in a particular skill, to master a profession, experience different cultures or as for me – simply to add value to people.

If you are inclined to adopt a seriously meaningful telos, one that may even transcend life, one with a purpose that is bigger than the individual or its realisation will exceed even a prolonged life, then you have cause indeed to direct yourself well, believe in what you do and enjoy the mission. Do we expect to achieve an objective if we don't know what it is - for all of us, shouldn't finding meaning and living our telos be our first and most critical task?

CHAPTER 12

Leadership

"To impress men's minds with a doctrine of development, will lead them in all honour to their ancestors to continue the progressive work of past ages, to continue it the more vigorously because light has increased in the world, and where barbaric hordes groped blindly, cultured men can often move onward with clear view."

(Edward Burnett Taylor)

The previous chapters have covered the rapidly changing technological world we are living in and have indicated that the information fuelled meteor has impacted the globe and sparked a new global era. There has been considerable comment about self-organisation, flat structures, no management and so on as the future of organisation and corporate structures. Such sentiments are ill informed.

New, exponential organisations will consist of its core members, a community of stakeholders (customers, supporters, partners, suppliers) and staff on demand. This organisation will interact with the community (everyone else) and as Salim Ismail states, in *Exponential Organisations*, "The more open the community, though, the more traditional and best-practice-oriented the leadership model has to be".

He explains that "You need strong leadership to manage the community, because although there are no employees, people still have responsibilities and need to be held accountable for their performance".

There has been an emerging resistance to words such as command or control in management. Some believe that such words are becoming inappropriate in the modern world. I recall writing a management document where I stated the 'management 101' principles of *plan, organise, control* and *lead* only to have the document come back to me with *control* deleted and replaced with *partner*. So fearful are senior managers of being seen as controlling; they are the same breed that insist that children not be tested, comparatively assessed or graded at school which ignores the very competitive nature of human beings and the competitive world such children will be joining.

There are fears, misconceptions and frankly ill-considered paradigms in abundance about a diminishing role for management and leadership. There is however no doubt that leaders will be in high demand. In liner, non-traditional, more unstructured organisations and absolutely in exponential organisations, leadership and influence will be as critical as it has always been; did Steve Jobs lead Apple? Does Warren Buffet influence investors?

Fortunately, leadership remains about influence and the fundamentals will not change. Rather, the Exponential Leader (ExL) will need to equip some modern skills, will become very entrepreneurial and community focused.

If you consider yourself not to be a leader then please understand that you are. I hope you will accept the fact that influence is leadership; that your daily interactions with people around you are the most fundamental manifestations of leadership. You are an influencer, you are a leader.

Traditional and Best-Practice-Oriented Leadership

The best practice leadership model, the stages of leadership development and some fundamental leadership constants I have adopted and am about to detail all come unashamedly from my friend Dr John C Maxwell [8]. John has written about 79 books (he's not even sure sometimes) on leadership and associated topics, was named the Top Leadership and Management Expert in the world by Inc Magazine (May, 2014), ahead of Seth Godin and Jack Welch and has been the consistent number one Leadership Guru on the Global Gurus research organisation site for six years now.

I have studied under John and am one of his certified team. I acknowledge that I know of no-one better to study and quote on leadership. Given that I have a military and corporate leadership background I thought that there wasn't a lot more that I needed to learn – until I discovered John's teachings. The global recognition of John's mastery is no accident. It reflects a lifetime of study and observation, countless investigations, interviews, discussions with managers and leaders at all levels and the adaptation of proven concepts subject to continual feedback and refinement. The 'laws of leadership', the need and imperatives for growth and the mindset to become a person of influence, with all that this entails, reflect what has been demonstrated to work in the real and very practical world.

A key characteristic of John's work is that it 'puts people first'; that is, his teachings stress the fact that to be a worthwhile influencer a leader needs to be 'of service'; to have a 'how can I help' and 'how can I add value' mindset.

John holds the view that although many are 'gifted' in leadership skills and clearly have natural leadership skills that leadership can absolutely be learned.

Before I discuss how the impacts of exponential technology will affect leadership into the very near future I want to address leadership constants – established wisdom that reflects human values, social interactions and simply characteristics, attitudes and habits that have been demonstrated to work for leaders to date.

I acknowledge that there are a number of recognised leadership gurus in the world. One such guru is Stephen R. Covey who wrote in the foreword of John's 10th anniversary edition of 'The 21 Irrefutable Laws of Leadership': "…communicators make the complex simple. Rather than an esoteric examination of leadership, this book is more like a foundational instruction manual". It is because of this eloquent simplicity, the art of making the complex simple, that I rely on John's work to highlight 'take-aways' on the very important subject of human influencing i.e. leadership.

Levels of leadership

To achieve mastery in any human endeavour requires staged development, growth. Those gifted musicians, Nobel laureates or cardiac surgeons weren't born into the world as they are now. They were born babies and their current expertise and giftedness is a result of staged, probably very focused and deliberate, development. Leadership is no different.

To define and to establish the concept that leadership is a progressive skill, one that is developed through stages, an ability that matures through mastery and the application of specific principles, I lean on John's 'levels of leadership':

Level 1: Position - People follow because they have to.

At this 'beginners' level, people follow you because they have to. In the workplace or a specific situation, at this level, the leader gets only minimal support; people will reluctantly follow instructions, usually with little enthusiasm, following directions simply because they have little choice. Initially, there may be resistance, there is little buy-in by the followers and generally minimal effort or contribution to desired outcomes. This is the level of the newly appointed supervisor, the first level of promotion, the appointed team leader; spokesperson of a syndicate at school or university and it could just as applicably be the newly appointed CEO of an international corporation. It describes the startup level in a contextual situation. This is the most ineffective form of leadership and any worthwhile leader will quickly advance from this. Unfortunately, and typically in non-performing organisations, level one leadership tends to be permanent.

Level 2: Permission - People follow because they want to.

Followers at this level have grown to like, know and trust their leader to the extent that they want to help. People will do better work, volunteer and give more energy to support the leader with whom they have made a connection. The leader has 'won-over' followers by various means ranging from forming friendships, being an example, displaying care for his charges, helping his followers, listening, consulting, providing guidance, being fair and decisive etc. The leader at this level is getting significantly more value out of his followers.

Level 3: Production - People follow because of what you have done for the organization.

Somewhat self-explanatory; Followers are drawn to success, to enacted purpose and evidenced results. The leader has established a reputation (he, she gets things done, cares, listens, takes things on the chin, stands up for us etc.) and his followers are starting to contribute, to perform cohesively and impacting on the achievement of objectives; positively affecting the bottom line. This is successful corporate leadership and most good managers / leaders will settle at this level.

Level 4: People Development - People follow because of what you have done for them personally.

This is an advanced level of leadership and one that defines the exceptional leader who has significantly invested in his followers. Leaders at this level have enabled their followers to grow and develop personally, helping them into their areas of strength, supporting and coaching or mentoring followers to the succession level. This is the powerful trust and loyalty level where followers are equals, not 'daunted by the boss' but partnering with him in a very productive and personal commitment to help each other reach objectives.

Level 5: Pinnacle - People follow because of who you are and what you represent.

The rarest situation where the leader has become a legend and reputation creates the 'know, like and trust' relationship by default. Such leaders are usually established through repeat successes, created wealth, influence, media profile, sporting prowess etc. Examples are people like Bill Gates, Steve Jobs, John Maxwell and Tony Robbins; for you - leaders in any field that you greatly admire.

Like most of us, if you examine your own level of leadership you should discover that your level may vary with different people; with some you might be at level 3 or 4, with others, for example with people that have just joined your organisation or social group, you might still be at level 1; in all cases your ability to lead, to influence them will be determined by that level of leadership. Simple by being aware of the 'level' with each of their followers, leaders are able to deliberately elevate the relationship and their effectiveness.

As I've stated before, leaders are not born natural leaders. True, many are born with certain talents or gifts; traits that may make them very social, communicative, and influential and they may be perceived as natural leaders. For the vast majority however, leadership results from learnt, adopted behaviours and value systems. Nobody becomes a leader overnight; all transition through the above levels and those with talent may find that transition both natural and easy whilst others need to consciously work at it.

The 21 Irrefutable Laws of Leadership

In the '21 Irrefutable Laws of Leadership' John Maxwell describes the fundamentals of good leadership. He writes "No matter where you are in the leadership process, know this: the greater the number of laws you learn, the better leader you will become. Each law is like a tool, ready to be picked up and used to help you achieve your dreams and add value to other people. Pick up even one, and you will become a better leader. Learn them all, and people will gladly follow you."

His very firm philosophical belief underpinning his teachings are well expressed by J. Mariah Brown[63] in her article *Servant Leadership: Leading the Way Through Servitude* where she states, "servant leadership is having the desire to not only lead, but to do so in an ethical manner' and that it means making decisions that will benefit those within the organization as well as the organization as a whole, and putting the wellbeing of the organization and its members before self".

The Law of:

1 The Lid

Leadership ability is the 'lid' that determines a person's effectiveness. The 'lid' represents the limitation of effectiveness based on leadership ability. The higher your ability to lead, the higher your lid; this law directly links the leader's ability to the ability of an organisation, those that are led, to achieve. It means that an organisation cannot (normally) outperform the abilities of its leader nor are higher leaders, those with a higher leadership 'lid' likely to stay in the employ of a lower ranked leader. On a scale of 1 to 10, how do you rate? Assuming you are an above average leader and score, say 7, how likely is it that you will retain staff that are 8s or 9s? How, if you were, say a 4, would your organisation achieve an 8 or 9 level of performance?

Clearly, not only must a leader be aware of his/her ability levels; a leader needs to continuously grow leadership abilities to maximise effectiveness and success potential.

2 Influence

"Leadership is influence – nothing more nothing less." (John C Maxwell).

This is such an elementary truth that it should need little explanation. I trust that you accept that leadership is not a quality that is automatically conferred by holding a management title, a particular position, being entrepreneurial, knowledgeable or pioneering. Rather, leadership depends on your ability to influence, even your online 'klout' score[64], and your influence will, to a great extent, depend on who you are (character), who you know (relationships), what you know (knowledge), what you feel (intuition), where you've been (experience), what you've done (past success) and what you can do (ability).

3 Process

Leaders need to grow; to be, or remain an effective leader, requires continual personal growth. Today's rapidly changing world, now more than ever before, has no room for the ill-informed, unaware leader. The law of 'process' requires the leader to grow through deliberate, planned, day to day habitual learning. As John Maxwell puts it "leadership is developed daily, not in a day". In his book *'The 15 Invaluable Laws of Growth'* he teaches the values of:

- Intentionality - growth doesn't just happen
- Awareness - know yourself to grow yourself
- The Mirror - see value in yourself to add value to others
- Reflection - learning to pause allows growth to catch up with you
- Consistency - motivation gets you going – discipline keeps you going
- Environment - growth thrives in conducive surroundings
- Design - to maximize growth develop strategies
- Pain - good management of bad experiences leads to great growth
- The Ladder - character growth determines the height of your personal growth
- The Rubber Band - growth stops when you lose the tension between where you are and where you could be

- Trade-Offs - you have to give up to grow up
- Curiosity - growth is stimulated by asking why?
- Modelling - it's hard to improve when you have no one but yourself to follow
- Expansion - growth increases your capacity
- Contribution - growing yourself enables you to grow others

A formidable list of what needs to be considered to develop and adhere to a continual process of self-improvement.

4 Navigation

"A leaders is one who sees more than others see, who sees further than others see, and who sees before others do"

The law of navigation requires the leader to chart the course for his followers, his organisation or group. This is applied leadership. The leader draws on all his abilities; all those human skills conferred by the thinking brain, the neo-cortex enabled ability to plan, conceptualise and perceive, with the aspirational goals of the organisation or group. Navigating a course of action, a strategy and its operational implementation is a qualitative balance to turn a vision or an objective into a reality. Simply put – leading the action to achieve an objective.

5 Addition

Is it somewhat counterintuitive to hold that leadership requires adding value to others? It is nevertheless a fundamental law of leadership; it's the previously described key element of the fourth level of leadership. To gain the highest level of support, people need to know that you not only value them but they also need to be fully aware that you have invested in them, that you have added value. It is perhaps more intuitive to believe that for certain situations, short term objectives, responses to unforeseen events or simply for convenience, this law doesn't apply. But it does; it is in those specific instances that adding value, immediate and consequential, becomes critical. For level four leaders, adding value to individuals is as critical as adding value to the organisation.

6 Solid Ground

The often used phrase 'know, like and trust' has been generally accepted as a fundamental condition of sustainable personal or commercial relationships. To a leader, the most significant of those three words is trust. Clearly, people will not follow someone they do not trust. Trust is a feeling that others determine. The leader can only achieve trust by consistently dealing honestly, authentically and fairly with everyone. I say everyone, rather than just followers, because in our hyper-connected society – ahm, we're all connected!

A leader's character determines his trustworthiness (as stated above - character is a key influence determinant) and trust determines how solid the ground beneath the leader is; how committed followers are to the leader.

7 Respect

"Respect yourself and others will respect you." (Confucius, Sayings of Confucius)

The law of respect states that people naturally follow leaders stronger than themselves. There can be little doubt that a strong leader needs to respect himself first. This means a disciplined and constant embodiment and representation of values; living their higher purpose - discussed previously.

Respect requires that the leader has leadership ability, displays respect for others, has the courage to act, is accomplished, loyal and adds value to others. As John Maxwell says: "When people respect you as a person, they admire you. When they respect you as a friend, they love you. When they respect you as a leader, they follow you."

8 Intuition

John Maxwell describes this law, the law of intuition, the most difficult to teach. This is not surprising given the text book definition of intuition is the 'ability to understand something instinctively, without the need for conscious reasoning'; that is 'knowing or understanding something without reasoning or proof'. Often, intuition is described as instinct and clearly one can't learn instinct.

Intuitive leadership however suggests that everything a leader perceives, considers, reacts to and does, is done with the benefit of a trained intuition. It is a subconscious adherence to everything that has been experienced and learnt and added to basic instinct. It is the result of the adoption of values and lessons learnt to effectively see, understand, and act almost automatically – intuitively, with a leadership bias. A leader's intuition can't be learnt as a discipline; rather it embodies deeply felt beliefs that the mind has accepted as fundamental truths; so much so that it is perceived as being without conscious reasoning. For a successful leader that means the genuine knowledge and acceptance of the values that determine leadership ability as described in 'influence" above.

9 Magnetism

Personal magnetism enables you to influence and therefore lead others. Magnetism draws people to you according to how interesting, how attractive, alluring, fascinating and charming they find you. In the leadership context people are drawn not only to these personal attractors but also to your evidenced leadership skills. Followers will be drawn to you, if and only if, they perceive you to be a good leader. Again that is determined by your display of leadership characteristics; your character being most important.

Not surprisingly, who you are, how you 'come across', initially determines who you attract. If you are all that you are perceived to be, that is, truly reflecting who you are, then such attraction will be more permanent (the be yourself argument). A practical application of the law of magnetism is to analyse what sort of person and leader you really are by examining your followers and people you associate with. What are they like? People are attracted to others like themselves; the qualities of your followers reflect who you are and are a reliable indicator of what you are doing right and what you might change to attract people with the desirable qualities that you are looking for. Manifest in yourself the qualities you want to see in your followers.

10 Connection

This law can be summed up with the words of: "Leaders touch a heart before they ask for a hand" (John Maxwell). Communicating is what we all do, connecting is what we need to do to be directly relevant in any leadership, mentoring, coaching or speaking role. Connecting with people requires an emotional attachment. This can be in the form of respect, mutual understanding, an aware commonality, shared attitudes or values; basically anything that links people. A simple example is meeting up with a fellow countryman in a different nation, recognising the common origin and making an immediate connection. People are more likely to follow a leader with whom the share a bond. The stronger this bond, this connection is, the stronger the desire to follow and to contribute to the leader's goals.

11 The Inner Circle

A leader's effectiveness and potential to succeed is determined by the quality of the team. The inner circle, those closest to the leader, is critical and needs to not only connect with their leader; they must align with objectives, compensate for any weaknesses and add value.

Representing the people that have most influence over the leader, the inner circle makes or breaks the leader. Leaders require support. Choosing the right people to be part of that inner circle not only enables good leadership it provides for succession and continuity, helps to keep the leader grounded and multiplies the leader's reach and effectiveness.

12 Empowerment

This is one of my favourites. I've witnessed, unfortunately all too often, the 'buffoon leader' in action. This is the appointed senior leader often found in government departments who has, by virtue of an 'old boy' or drinking buddy connection, been floated to the top. They are the insecure, often quite stupid bosses (not leaders) that survive in their tortured position only because they naturally surround themselves with like characters (magnetism, inner circle, connection etc.) that do not question, exact no accounting and contribute only to the status quo, the situation that has brought them into the circle. These are the weak bosses that cannot delegate beyond their circle; they protect their insecurity by holding the reins of power very tightly.

A negative example I know but one that is more evident than the strong leader delegating and empowering others. Empowering others, as in level four leadership, carries risk; the risk that better solutions, innovation, thoughts and even objections, might come from lower position holders; the very aim of empowerment, to get the best out of subordinates and to facilitate feedback, is what the strong leader wants and the buffoon shuns.

13 the Picture

Simple but important – set the example! People do what people see, people copy what works and what is acceptable.

14 Buy-In

Most people will buy into an objective or a corporate vision only if they accept and 'buy-into' the leader. Most of us have listened to vision, dream or objective presentations and yet not been moved into a 'rolling up our sleeves frenzy' keen to get on with making that happen. Why? We didn't believe the message, didn't support the leader, thought it was, well BS, or simply didn't care enough for many reasons. Certainly we didn't buy-in sufficiently to make that leader's objectives a priority for ourselves.

The key here is again simple, be an effective leader first before asking people to follow.

15 Victory

Being a good leader does not guarantee victory. There are numerous examples, particularly during military conflicts, of skilled and capable leaders who did not win. An example is a high risk rescue attempt where the value of the operation is so high that an attempt is warranted against overwhelming odds. Winning or losing in a non-extreme situation, in a personal, social or corporate sense is all about achieving objectives and here a leader has great influence.

To win, a leader must be absolutely clear about what is to be achieved, what it will take in resources and be totally committed. Further, that vision, understanding and commitment must be communicated (achieve total buy-in) to the team. The leader's conviction that the aim is achievable, his clear purpose and determination to achieve it must result in a team unity of purpose. The law of victory demands all that the leader is capable of bringing to the organisation. It's the culmination of all leadership abilities and skills. Leading a team to achieve is the purpose of leadership.

16 Momentum

The law of momentum is important because as John Maxwell states: "If you've got all the passion, tools, and people you need to fulfil a great vision, yet you can't seem to get your organisation moving and going in the right direction, you're dead in the water as a leader."

You may have some experience with, or awareness of, organisations where moral is poor, enthusiasm is lacking; there is a history of losing, little energy and a perception of hopelessness. Such unmotivated cultures exist in many organisations and they can be described as having no go forward momentum. Momentum builds when even small achievements are recognised; when incentive is provided and successes start to feed on themselves and become habitual.

Leaders can create momentum by eliminating negatives. This might require repositioning or changing staff, allocating resources, even over resourcing to ensure a minor win. The leader must change the mindset and engender renewed belief in the team's capabilities by reinforcing successes however small, and excising negatives. When the team perceives improvement they will perform better. When such improvements change group mindset to one that accepts that achievement is possible, winning, achieving becomes the new expectation; belief is restored and positive, go forward momentum pervades the organisation. Momentum builds on successes and makes this a corporate habit.

17 Priorities

The staff is busy doing exactly what? What are people actually doing and how is this directed. Human nature is to do what is comfortable, pleasurable and easier first. Difficult tasks are left for later and whilst they may be priority tasks they just don't get the attention unless priorities are set and enforced.

Setting priorities is a leader's responsibility and reflects his management skills. Priorities flow from the leader's appreciation of the task based on the plan, organise, control and lead process. Having determined the task, the order, the priority, the leader needs to ensure that priorities are implemented.

The above is quite logical isn't it? But the law of priorities isn't about what a leader's staff is doing, it's about what the leader is doing to lead and prioritise his own actions, development and direction. This concept, so easily agreed as vital for an organisation is nevertheless mostly ignored by individuals. Leaders constantly evaluate their own priorities. Time and energy limited, the leader must self-direct and be disciplined in its execution. Being busy is not good enough. A high rate of activity is not an indicator of appropriate action; it is a deception in much the same way as mental noise is not an indicator of deliberate thought.

Understanding what demands your attention and its relative priority is critical. An awareness and application of the 'Pareto principle'[65] can greatly assist in clarifying the amount of effort any one issue deserves.

18 Sacrifice

The law of sacrifice acknowledges that there is a price to be paid if you want to grow as a leader. Living the life of a leader means that you need to subordinate your self-interest to those of the group you are leading – and that's a tall order. Leadership assumes a significant responsibility over others and the higher the level of leadership the greater these responsibilities become. We all value our own time and the pursuit of activities and things that make us happy. The leader finds a balance and sacrifices personal desires for the good of the team.

History is filled with examples, some inspirational, others tragic, that the commitment to others, to a noble goal, does exact a price. Every day, we see leaders working long hours; they are sacrificing one of humanities most precious gifts – time. They are giving up family time, personal recreational time and time to do whatever fulfils them. Giving up their own time to better lead others, often with the knowledge that if they lead well and increase their leadership domain, they will have to give up more.

Leaders need to understand what they are foregoing and monitor the desire or need to lead against values they hold; what sacrifices are they prepared or not prepared to make?

19 Timing

'Timing is everything' is a phrase you've probably heard often and of course it is not. The wrong action, regardless of 'when', is still wrong. The right action at the wrong time is equally not good. Obviously the right action at the right time is what leads to success.

Leaders use a range of determinants to establish the right time. Drawing on their acute understanding of context, the situational appreciation, they decisively, based on their experience and intuition, execute the right action at the calculated best time. Simple – just master those qualities and you'll get the timing right.

20 Explosive Growth

In the law of explosive growth, John Maxwell holds that 'to add growth, lead followers – to multiply, lead leaders'. The law requires an acknowledgement that there needs to be a change of focus; from followers to leaders. As described earlier, at the fourth leadership level, the leader develops the leaders on his team. By elevating and enabling those subordinate leaders to become exceptional leaders, openly encouraging succession ability, the leader causes the organisation to grow exponentially.

The focus implications are that leaders who attract followers: need to be needed, develop the bottom twenty percent, focus on weaknesses, treat everyone equally, spend time with others and grow the organisation by addition. In contrast, the leaders that develop leaders: want to be succeeded, develop the top twenty percent, focus on strengths, treat individuals differently, invest time in others and grow the organisation by multiplication.

The followers are however not neglected. Sustainable 'explosive growth' requires everyone to be brought along, so whilst the leader is focusing on developing the leaders below him, his leaders, are leading the followers and identifying emerging leaders themselves.

21 Legacy

What will be written on your tombstone? Hopefully something more than 'I told you I was sick'. You may not particularly care and certainly that's not all that important but it is a good way to think about your 'Telos', your higher purpose.

Why and to what end did you lead? If your leadership was focused on goals exceeding your own mortality then did you enable succession and is your purpose still being pursued by capable leaders? As I detailed in the previous section of this book, there is no greater value than to identify and live your higher purpose. If that involves influencing (leadership) then an eye to your legacy can not only keep you focused and motivated, but it will guide you to be a better, more thoughtful leader.

So they're the 21 irrefutable laws of leadership. I need to make it quite clear that in his book, *'The 21 Irrefutable Laws of Leadership'*, John Maxwell teaches these laws as only their creator can. My mindset is that of a student seeking to understand the master; I'm not teaching these laws; I'm interpreting and stating them because they are so fundamental to effective influencing. It is the ability to influence, in an exponentially developing world that I want to focus on. What, if anything is changing? Do the laws hold or are they being challenged?

The Exponential Leader

Numerous surveys, interviews, discussions with and among the world's top CEOs all share a critical acknowledgement that leaders, managers today (and the organisations they lead), are ill equipped to cope with the ever more complex requirements of leadership. In previous chapters we've looked at the current global environment, exponential technologies and we can therefore appreciate the looming complexities brought about by these technologies and the resulting social and economic impacts.

There is clearly a power shift away from the post WW2 victor dominated world to one dominated by the new, if re-emerging, world. This is a type of global democratisation where the most populous nations are determining their own values, directions and walking away from past hegemonistic paradigms.

This complex world will not settle into any steady, predictable state for decades to come. Our time is as volatile as it is exiting and it is pushing our leaders and our organisations to and beyond limits. The challenges facing leaders today are an entrée, the feast of confusion yet to come will disrupt mercilessly. This is actually widely acknowledged.

In the opening chapter I mentioned Singularity University's 'Innovation Partners Programme' and that 75 percent of C-Level executives who attend admitting to having no previous awareness of the technologies involved. Imagine returning to your CEO or your board and admitting to them that the information inspired meteor had landed some time ago, that there was significant almost revolutionary change happening around the world but that you didn't know! Explain that in all likelihood there is someone in a garage, or there's a start-up somewhere in the world, that is about to announce a breakthrough technology that will dramatically challenge your product or service and that your organisation is about to experience its very own Kodak moment.

However, now that you know, you will immediately spring into action, fundamentally change your company structure, innovate, adopt new methods of operating, engage with the community, outsource the majority of the staff, do away with hierarchical structures and job titles, sell all but critical assets and a few other vital things that will hopefully allow you to survive the imminent disruption that numerous start-ups are currently working on.

Is the above an unlikely or farfetched situation? Not at all - such discussions are becoming frequent in board rooms around the world. If you're a leader with executive responsibilities – are you ready to lead you organisation through this technology revolution?

Exponential technology has been with us for some time. The well-known Kodak experience along with such notable technology disrupted industries such the newspaper publishing, taxi companies and media have all experienced technology based disruption and there is more to follow. Many leaders however are well aware of disruptive technology and unsurprisingly some of the largest Chinese companies are at the forefront.

Xiaomi Tech (produces lower end android smart phones) and Haier (appliance manufacturer) are outstanding examples of exponential organisations operating today. Not only do they engage very effectively with their communities, they have adopted corporate cultures, management and operating practices (even profit sharing and people empowerment) one would not normally associate with a communist state.

What then is the C-Level executive's or leader's excuse for not knowing or understanding the technological impacts influencing everything at a rapidly increasing rate?

Unfortunately, 'consciously' working on developing leadership skills takes time and effort. Mastery requires effort and time. The exponential future that is approaching will severely test our ability to master anything; as opposed to having instant information but little in-depth appreciation. Certainly, time will be short for leaders in any commercial or governing position and there will be significant self-skilling pressures on would be leaders and a consequential educational liability to be accommodated.

Our world is being described as volatile, uncertain, complex and ambiguous. (These words have been borrowed from the American Military that coined the acronym VUCA). 'Volatile' is apt because of the rapid pace, scale and wide effects of change; 'uncertain' – yes, because the future remains as unpredictable as ever but with the added fact that change can now have massive, immediate and global impacts. More 'complex' – clearly; most leaders describe their environments as stressed due to information overload, hyper interconnectedness, failing organisational boundaries, disruptive technological threat potentials and practices, time poor and flooded with change. They acknowledge a certain redundancy of the 'status quo' and feel themselves part of a world that is simply not equipped to manage or govern today and has little or no plan to manage tomorrow.

'Ambiguous' is appropriate because of the previous three characteristics; the future is not clear and there is little confidence that we can predict cause and effect, technological or geopolitical outcomes.

I think it's fair to conclude that management in general, leaders, are stressed, confused, overwhelmed and generally feeling inadequate to meet demands. This is of course hardly surprising. We've trained and developed individual leaders to be competent against a fairly static list of skills and knowledge. Whilst many fall short (just evaluate leaders you know against the above 21 laws of leadership) we at least knew the competencies they should possess.

But our training and development hasn't changed much since the industrial revolution and we are seeing that on-the-job experience, training, coaching and mentoring is no longer adequate. In the past, such training, the development of leadership skills, has been the responsibility of others rather than the individual concerned. It was the human resource department, the manager, the trainer or an institution that had analysed the required or desirable competencies and helped to ensure that the leader learnt and developed these accordingly. We now find that whilst such development remains necessary it is no longer sufficient.

The skills, attributes and abilities most commonly cited as lacking and of key importance in the future include self-awareness, adaptability, collaboration, network and strategic thinking, creativity, change development and management. Clearly what will be needed are competencies that ensure adaptability, the ability and commitment to learn continuously and above all to be able to thrive in an ambiguous world.

Organisational leadership needs to transform from the current top-down, hierarchical and organisation centric model to a two-way, collaborative and network centric environment. Future leaders will need more complex and adaptive thinking skills. Additional to the traditional skills and knowledge, leaders will need to develop new mind-sets and they will need to know 'how to' and 'to implement' their own development. There simply aren't the courses available. The volatile, complex, uncertain and ambiguous nature of the future leadership environment has not yet revealed the definable and teachable changes such mind development requires.

Horizontal and Vertical Development

The prime, first step is that leaders need to reclaim ownership for their own development; just as espoused in the previous chapter. The required development is being defined as both horizontal and vertical development.

Horizontal development is the traditional skill and competency based development that shapes leaders' behaviours and performance. Technical by nature, it is development that has been observed and defined as core competencies that a leader needs to possess and there are individuals, institutions and texts that seek to help the leader in developing those skills and abilities. A good way to think of this is likening it to a personal computer. You expand its capacity and capability by adding software. The situation we have now is that we have too much software; the computer is running too slow, the operating system cannot cope well with the computational load and it ran out of spare memory years ago.

Vertical development seeks to address this. In the 'computer example' it's like replacing and seriously upgrading the hardware; replacing the old

PC with the latest, fastest, multiple core processor driven machine and ensuring that its interfacing is connected, user friendly and massively capable. Vertical development seeks to improve how the leader makes sense of the world, its directions, opportunities and threats and to understand that world in great detail, inclusively, in context and all of its complexity. This development seeks to alter the leaders' mind-set to develop better capacities of thinking.

In vertical development, successive levels require learning, complex problem solving, direction setting and a change accommodating mind set. Which each step up the vertical ladder, the leader can learn more, adapt faster and solve more complex problems simply because they see and understand more, they are making sense, connecting more dots and getting better at strategic and contextual thinking. Ultimately they don't just manage change, they drive it. Leaders used to be assessed on the questions they asked. Future leaders will be assessed on the answers they provide and more importantly, on how they got to those answers.

Assuming our typical leader has a high degree of horizontal development, knows and lives by the 21 fundamental laws of leadership and is at vertical development level one. He would typically be a conformist. He would conform to social paradigms, ideology and norms with a strong desire to meet expectations and to fit-in. At this level, his thinking is dominated by approval and compliance.

At the next level, vertical development level two, this leader would display a fair degree of independent thought based on awareness and understanding; seeing the world in context and institutions as they are rather than what we would like them to be and to be comfortable with perceived imperfections and resulting uncertainties. This leader brings a conscious deliberateness to actions and changes, accepts his telos and ideologies and no longer accepts belief systems at face value. This leader is shaping opinion and becoming a thought leader.

At the third vertical level of development the leader has established firm ideological frameworks but maintains a flexible mind-set. He can hold opposing ideas equally valid, he can be objective about perceptions, is open minded, inclusively tolerant and confident with difference. This person can accommodate different belief systems and readily adapt to changes in environments, forms and functions. This person is a thought leader with a level of wisdom that transcends personal focus and purpose.

Critical to any advance in vertical development is the 'how'. What exactly should we do in order to advance; how do you get a bigger mind? Well you could wait for technology to literally enhance your neocortex. There might well be an AI 'help you think' application on the horizon that you could use to solve problems but we actually need the process; we need to understand and assimilate, make sense of a larger world. And the way to get there is actually relatively simple. The first step of course is to want to!

At the beginning of this book I wrote about my awakening. The opening of my eyes when I was confronted by what is and what will shortly be possible with exponential technologies. That's the first step – the wake up step. It really is similar to a mid-life crisis. Look around you; look at your life, your nation and the world objectively. Pretend you're an alien and draw some conclusions about what you see. Look at your own purpose, the purpose of countries. Realise that everything is changing, in a state of flux and admit that it has always been so. Nothing is permanent; there are no guarantees and no certainties. This should not depress you for long; it should be a freeing experience that allows you to make some sense of the world and open your mind to a multitude of options and possibilities. Above all it should release you to accept different paradigms; to let go of self-limiting labels and beliefs.

The second step builds on the first. Seeing and making sense of the world should lead you to examining old beliefs and acknowledging the value of difference and diversity. It should lead to you defining your telos; what's worth doing for you and what will you sacrifice to get it done? Living your telos, linking that to an organisational MTP will radically shift the way you live, work and play.

The final task is to grow. Learn more, think more, understand better and excise limiting behaviours and ideas. This growth might lead you to re-examine who and what is influencing you; you might change your friends and even partners; it matters little when you have your very own purpose. You will accommodate new ideas and as they take hold you might drop others. The key to growth is to stay interested and to keep learning. Growing is a lifelong activity; the concept that an adult is 'grown-up' is not valid.

So, three steps to advance on the vertical development ladder - I know it's not that easy but I know that many have embarked on this journey already. In writing this book, I deliberately structured it to lead you to the next vertical level.

I used the example of my own awakening so as not to offend you. Better to say I've been asleep (true in any case) then to suggest you might be – no? That whole chapter took your mind, perhaps subconsciously, on a journey of discovery. Presenting exponential technologies, sometimes in too much detail was meant to show the complexities of innovations and to raise an awareness of the possible; what technology can bring, the opportunity of abundance.

The short but critical examination of the world today, the history, the institutions and the changing world order inviteed you to see the world with some clarity. You needed to make sense of the complex world and observe that it does not reflect a utopian 'god managed' kingdom but one that is constantly changing and challenging.

The chapters on the near future and self-direction were to guide you to accept that there is some urgency and need to grow vertically. The concept of the new leader being self-developing was introduced in the leading yourself arguments.

So if you stuck to the plot then you have as a minimum entered the second level of vertical development. You may not feel particularly enlightened or different but let me assure you that you have changed. By simply reading and hopefully thinking about what you're reading, your mind has been broadened; the mind knows what it knows, assimilates what you have been interested in and exposed to, and it expands accordingly.

There are theorists and academics that believe that the era of 'heroic leadership' is coming to an end. They hypothesise that a new 'collective' leadership model will replace the individual leader. Some believe and are pondering ways to spread leadership throughout an organisation and to democratise leadership; basically pushing the concept that leadership will no longer reside in individual managers, leaders, but that it will be spread throughout the network. Wrong, totally wrong and missing the point.

Not only is leadership a human constant, there will always be influencers among us. Until the world changes the rules of might, economic power and the human desire to 'have, know and experience more of everything', this will not change. There will remain the human competitive imperative that will seek to 'win' and in order to win, groups, societies and nations will look for, and turn to leaders to get it done.

They contend that a networked collaboration in information sharing, planning, cross influencing and decision making is required and they are spot-on; it's called inclusive, collaborative or consensus leadership and this is well understood by today's leaders. No new model is required. What is desperately needed is an almost religious adherence to the 21 fundamental laws of leadership and the application of the new mind-set of the vertically elevated leader. That leader is an exponential leader.

If I had to detail the one overarching characteristic that defines exponential leaders it would be that they mine and utilize the cognitive power of their employees and associates. Employees are the cognitive reserve of any organisation. Exponential leaders consider these people as a part of their own mind but sufficiently detached independent thinkers, self-organising, alternatively skilled and knowledgeable, analysts, problem solvers and innovators. If these employees align with the corporate MTP and understand and share the leader's objectives, they will become an adjunct intelligence and immeasurably enhance the leader's neocortex, his intellectual ability to deal with challenges. Exponential leaders know and live this and that is why they are the people that will lead us into the future.

In exponential organisations, this cognitive resource expands beyond the reach of employees and includes the online crowd. This enhances the organisation's mind and capabilities exponentially. What can be achieved when the cognitive resource of the hyper connected world is tapped?

So what's changed?

Are the 21 irrefutable laws of leadership enough? Can we just include the above horizontal and vertical developments to complete the leader's training syllabus? No.

I detailed the above laws and development needs because they are critical leadership skills. They will continue to be that but the future will demand far more. I am convinced that today's demands on leaders are already beyond any level of comfort. Most leaders know this. There is simply an overload of cognitive demands, too much information and an expectance of immediacy in all responses. Gone is the time for contemplation, assimilation and deep understanding.

Our ability to lead and to develop subordinates, to develop the next generation of leaders and to invest in people generally is rapidly disappearing. How do you lead a team that you've had no chance to get to know or to develop? How do you connect (as in effective communication) with people that you only interact with online? How do you know, like and trust someone you have only met 'electronically'? These are challenges for the future leader.

The teachings of gurus like John Maxwell will always be fundamental in developing leaders; they have served the world defined by industry and proven their worth; unfortunately they will not be enough. The information age is bringing challenges that have not yet been defined but we can well anticipate these:

> **Time** - there isn't enough time (although we all have all the time there is). Gone are the days of investing time in staff. This is the logical result of allowing industry to dictate

requirements to teaching institutions that have been under relentless pressure to produce 'work ready' employees. This has gradually freed industry from mentoring and training staff. Today, there is no appetite and no time to train or develop – want the job then perform from day one.

Time - I've already dealt with the demands on personal thinking time; this is obvious and it's taking its toll on leaders everywhere. Managers at all levels are finding it difficult to cope; there is no time to read and understand the multitude of papers, reports, and general correspondence and even emails. The result is a level of power browsing that only looks for the most important. The result is shallow understanding, growing uncertainty and the collapse of confidence.

Access – this is a new paradigm of the lean organisation of the future. Given that in-house human resources will be maintained at minimums, talent, support and assistance will be sourced on an as required basis. This leaves little room for knowing who you're dealing with even if you can source and access your requirements. This problem is already becoming apparent in many start-ups. They have difficulty in accessing required skills and when they do they have little confidence in priority of effort or successful outcomes.

Loyalty – This has long been taken for granted for permanent employees but is becoming a problem to the modern organisations' leaders. When you employ on a needs basis, where is that employee's loyalty? Whilst there is the fair expectation that work will be completed as contracted there is the understanding that as soon as the task is awarded the 'employee's' attention will be drawn to the next opportunity. Anyone who has ever worked as a consultant will know this.

The above are just a few of the obvious challenges to be met by leaders. Like everything however, this will be greatly influenced by the environment and that is likely to alter significantly.

Just as personal time has all but vanished so is time for any commercial development, process or advantage. Being first to market will mean little; someone, somewhere in the hyper connected world will be copying and improving. Gone too are prolonged product development times. Products will be pushed out pre-completion; the crowd will be engaged in its development, refinement and to an extent prequalified as customers. It is also likely that time 'in business' will be product related with companies dissolving when limited objectives are met only to re-emerge with a new offering in the future.

Massive change will affect us all. Those in positions of leadership are in for a tough time. They will need to be extremely knowledgeable. This will be against the trend of 'no need to know – look it up, Google it when you need to know' that is pervading society now.

Our future leaders will need to know and apply the leadership lessons of the past and they will be required to adapt and maintain intellectual flexibility to lead in a world that we cannot today understand.

CHAPTER 13

Governance

In Part 1, I introduced three pillars to assist our understanding of the future.

The first was 'being human'. Understanding our humanity, our embodiment, our fragilities and social clustering defines us as the sentient beings that dominate our planet. We share our limited 'lebensraum' with more than 7.4 billion people; each with their unique existential reality and equally unique telos. I've described the global context that we experience and our rapidly changing geopolitical and environmental landscapes. Whilst many may not agree with my observations we should however have no difficulty recognising that significant changes are taking place.

The very human abilities and characteristics that have allowed our species to thrive, grow in number and dominate all other lifeforms are continuing to evolve us as a species. This evolution, now fuelled by technology is happening at breakneck speed. Exponential technologies, the second pillar of understanding, are enabling greater and faster diversification; a diversification that represents inequality, advantage, disruption and opportunity. By far the most significant aspect of technology is that it is allowing us to transcend the limitations of being human. Our species is transitioning from slow, gradual and controlled biological evolution to a much more immediate technological evolution. This transition will lead to a greater diversification of humanity; this will be spectacular in scope and speed when strong AI is realised.

We can do little about our underlying humanity. We can equally do little to halt or control exponential technologies. There are many great minds considering the fact that technologies are being developed with 'no plan, no control and no brakes' and they have sought ways to control or even stop various technologies. It is now clear that only a global totalitarian regime, however undesirable that may be, would have the best chance to control or halt technological developments. It is however acknowledged that if that were to be the case then research would simply be done 'underground'.

For better or for worse, we cannot stop the relentless progress of technology (no agreement that we should in any case). We cannot yet change humanity – what do we do?

Technology, people and events are leading us into a future that cannot be imagined or understood today. How do we manage this? Who or what will lead us into the future?

Understanding and dealing with the implications of technologies in the global and human context; the 'so what', 'why it matters' and 'what if', requires wisdom and its application to our management of humanity. This, the third pillar of understanding, is our grandest and most difficult challenge. It is the challenge that if not met, renders all else futile and will most probably end the human story.

Why now?

Whatever your take on the political, social and economic state of the world is, massive change is coming and it is severely overdue. Right now, today, with all the technological advances available to us we have failed to bring about any unified, peaceful environment. We cannot adequately feed, house or provide basic health care to billions of our fellow human beings. Whilst we may prefer to think that all things are right and beautiful and that humanity is fundamentally good; I believe that an accurate reflection of our history argues the opposite.

Our global conduct has been characterised by the use 'force' and 'fraud' strategies. Force includes all those human actions that seek to dominate either by applied, indirect or the threat of military action. It's the sort of global bullying that has been the practice of the developed world. Any nation once colonised, conquered, annexed, pillaged, enslaved, politically liberated by external forces or simply subordinated is testimony to this. Fraud is that despicable moneylender behaviour that is evidenced by the majority of our global financial institutions.

Our 'greatest' banks (IMF, WB and the Bank of International Settlements) have served their hegemonistic masters and themselves well. They have profited from the, force established, superior positions they create and maintain. As Thomas Jefferson pointed out long ago "I believe that the banking institutions are more dangerous to our liberties than standing armies..."

The current era of force and fraud is drawing to a close. Whether this happens in a deliberate and orderly manner or is the result of violent rebellion in our exemplary and failing democratic nations (the west) is yet to be discovered. The emerging kingdoms, principally China, Russia, India and their new cooperative allies are already significantly impacting the status quo. No longer is the world hostage to Washington or Wall Street. The new order is offering alternatives; alternate banking, infrastructure support, trade opportunities and peace. It is doing all of this without threat or impositions; no regime change, no 'Greek' rescues, no demands to change this or that.

So why is governance important now? There are the obvious 'good management' factors but what makes change critical now is the convergence of a number of significant developments. These include:

> **Transparency** – the result of our hyper-connectedness and the instant nature of information dissemination; basically, lies and deceptions no longer work as well as they did; motives are critically examined, exposed and challenged. This is becoming evident also in the voicing of objections

to past 'rape and pillage' conquests, colonisation and invasions. More and more nations are finding their voices and correcting the history books that have been written by victors. We even see examples today where nations are pointing to the great collections of artefacts in western museums and galleries and correctly asserting that these are plundered goods. For decades, the west has upheld its self-arrogated right to dominate and to dictate. This era has come to an end; existing powers however have not yet understood this. Their words espousing their values, fairness and stewardship are hollow and seen as such.

Lethality – today's weapons are far more deadly than ever before and they are easily obtainable. From 'traditional' weaponry to remote delivery systems and sophisticated weapon platforms our ability to cause mass harm to life and environment has never been so great. Include the new weaponry of cyberwarfare, chemical and biological agents and the threat profile becomes truly concerning. Technology and the ready availability of these weapons have changed the security task to one that has become fundamentally reactive.

The Cornered Bear – It is reasonable to expect that like a wounded and cornered animal, the old powers will fight to retain their self-arrogated positions. When nations find themselves in 'no win' situations they may believe that conflict, even if significantly self-destructive, is a viable option; the 'nothing to lose' motive is powerful. Consider a nation racked with violence, inequality, civil disobedience and a bankrupt administration – would they instigate conflict to wipe debt, to change the status quo, to focus and divert the population, to unify it against a common threat? Don't dismiss this – this is a real possibility and could be used by our champions of pretend democracies.

Hypocrisy – This is the most arrogant quality of our global hegemonistic pretenders. How can any nation maintain and continue to develop nuclear weapons whilst at the same time use coercive force to stop any other nation from developing the same weapons? So just to be clear, we can do what we want; you can't because we said so. As nations recognise alternatives to the dictates of Washington and New York they are gradually finding the courage and means to exert their sovereignty. This is evident in the resistance to various trade pacts the US is frantically trying to force on its vassals in the futile attempt to maintain advantage that has already been lost.

Technology – Developments in weaponry and disruptive technologies will accelerate the process of passing 'power' into the hands of many more players than what has been the case in the past and it may also render previously considered deterrent nuclear weapons ineffective. We will soon become aware of new weapon technologies (laser, photonic and electromagnetic pulse weapons are under development). These will be followed by nano-warfare systems (robotic insect swarms and micro weapons) and biological agents that will be capable of targeting specific subspecies (genetically identified). These new technologies will severely disrupt and change the balance of power. The technologies to bring once powerful nations to their knees and render them defenceless will emerge within a decade.

The Threat of War – Given the emerging social upheavals flowing from various causes like inequality, radicalisation, human dislocations and climate change the threat of war will escalate. Our human history evidences the rise and fall of civilisations. Whilst many believe that this is unlikely today they need only observe the very near future. Is it likely that China will go to war with the Philippines, Malaysia, Taiwan or Vietnam over the Spratly

Islands? No. Is it likely that Russia will invade the Baltic States? Of course not. These re-emerging powers have no comparable history of invasion or instigating war to America or Europe. It is far more likely that the US and its hapless lackeys will again cause conflicts as they have so incompetently and illegally done in the Middle East. They have learnt little from successive losses and intervention failures; only they can keep losing wars and still consider themselves superior. National and corporate economic benefits of the western arms industries are powerful incentives to ensure conflicts are sustained somewhere in the world. Global conflict is a real possibility; it is erroneously seen as a viable option by too many.

The above are all existential and growing risks. How can we mitigate these threats? I believe we have little option but to work towards global governance and thereby apply wisdom to our management of humanity.

The need for us to globally implement 'better' governance is growing in urgency. As empires start to collapse (principally Europe and the nations it spawned) many nations will experience social upheaval that will make any planned and deliberate transition almost impossible to implement.

This isn't soothsaying; it's fairly obvious to even casual observers of the growing tensions caused by an extensive menu of dissatisfactions throughout Europe and the US. Once the ability for controlled civil change is lost the only response is force. Most nations do actively plan for civil disobedience events and they rely heavily on their police and military to enforce law and order. That works only as long as the enforcement agencies remain loyal and are prepared to take up arms against their own people. Eventually, when confidence and trust in discredited political administrations and leaders is lost, then revolution is inevitable.

If you believe that this is not possible or likely in our modern and 'civilised' world then you've neither looked nor 'seen' history's evidence. Ethics and morality are not new concepts. Long before and after sages like Aristotle and Socrates philosophised about human traits and abilities to act violently and bring about the fall of civilisations have we done just that. There is no evidence that we are any more 'civilised' today.

It is clearly preferable that we acknowledge that things need to change. Global political ideologies are being shown up as fundamentally fraudulent [*you don't still believe in true democracy or global human rights – right?*]. Trade and commerce, the cornerstones of capitalism, are finally seen as the exploitative activities that they are and as the world runs out of ignorant, undeveloped and backward nations to subjugate and 'trade' with, these commercial activities will become unsustainable.

Technology will transfer 'knowhow' that will drastically reduce the physical exchange of goods. Consider how genetics and nanotechnology (see Chapter 9) will revolutionise the production of consumer goods. Consider also the future of abundant energy and the freeing of societies from grid infrastructure. What might the impacts on world trade be?

It is difficult to contemplate our near future but it is coming to pass. Most of us can't imagine life without work. Do we just lounge about, live in virtual realities? What will we do when there is very limited commercial business?

Change is coming; revolutionary change. The urgency is simply the choice between revolution 'mild' or 'wild'.

Global Ethics

The most critical need for the world's population is to adopt a form of unified consciousness; an awareness and acceptance by all of the fact that every human being has the right to be empowered to live a peaceful life with dignity and freedom. We talk a lot about the hyper connected world of computers but we need to take this to the next level i.e. we need to network intelligence leveraged in information. This new consciousness needs to facilitate new social structures where all can contribute their talents, learnt or natural, and contribute to the common good.

I invite you to dwell on the need for a global moral framework; a non-aggressive intelligent and ethical framework. Difficult to imagine but necessary given today's failed ideological state. There are many who believe that we are on the verge of the next big mass extinction event and that if things don't change the fate of the anaerobes awaits us too.

Today we have many of the world's most powerful nations still believing in the 'force and fraud' tools of achieving their self-serving objectives. This is a worry. With the rise of exponential technologies the tools of conflict, traditional and high technology weaponry, cyber, nuclear, biological and chemical warfare has the potential to be vastly more devastating than ever before. The information age has also equipped a plethora of organisations (criminal, terrorist, extremist and rouge nations) with the abilities to do harm at increasingly larger scale and with global consequences. Just as we are all connected and reliant on modern technology so are we vulnerable and increasingly incapable of surviving any length of time without it.

Globally effective, moral political debate is critically overdue; technology is marching on at an exponential rate and we are not in any way ready to govern the world it is transforming. That is the key issue. The case for global, non-aggressive governance is strengthened as the world exposes at increasing rates, continuing instances of compliance compulsion and the exertion of interference and unwanted influence (again 'force and fraud').

There are mounting examples of revelations such as national intelligence operations, specifically against allies, political interference, false flag and black operations that are a clear result of data leakage, information sharing and the connectedness of social media. The fact that these issues are being unearthed and aired is a 'blowback' that has become a form of social moral defence, at once exposing and disarming the perpetrators and revealing their naked and unethical conduct. Such moral defence has great potential to motivate and cause conflict particularly where large sections of populations experience injustice, inequality and perceive themselves to be indifferently governed.

The alarm bells should be ringing in every corner of the globe. Our ideological differences are not diminishing and the technologies being created now will only exacerbate and perpetuate our unsustainable global psychology. We are faced with an evolutionary irony of our own making. Having managed survival 101 we now need to survive each other and contain the technologies that can either bring about abundance but may just as likely lead to our extinction. For the first time in our history we are in a position to significantly take control of how we evolve.

Controlled or not, we will continue to evolve. To date, although our evolution has been spectacular in innovation and progress, it nevertheless evidences socially barbaric human behaviour; behaviour that is ethically and morally indecent. You might also acknowledge that our recent human practices are also ecologically unsustainable and that we are quickly reaching a stage where the survival of our species is in doubt. No, this isn't focusing on the negatives. It's simply stating the obvious.

Our collective task is to ensure that we manage our future evolution for the simple reason that to maintain sentient life on earth we must accept a degree of intentionality. The stakes are simply too high to leave evolution to chance, to nature; survival of the fittest is not a kind or humanistic approach. Nature and therefore natural evolution knows no 'sorry' for those that are left behind. We however have rightfully adopted values that make it unethical and immoral to leave anyone behind.

We are in the early stages of a technological revolution and what we need to survive is a global intellectual revolution. Whilst technology can indeed deliver all that it promises the most fundamental challenge is to ensure we adopt systems of governance that will mindfully usher in the application of exponential technologies; technologies that will radically change our lives. Trade and commerce, the value of goods, the global competition over resources, the administration of peace and the containment of borderless crime – it will all change. It is really a lot simpler to try to define what will not change and I'm assuming that the humane traits of our species, compassion, ethics and morals will not only remain but will manifest globally.

Mankind has achieved so much very rapidly. Now we need to share our innovations and it needs not be a zero sum gain. Leaders need to look seriously at humanity and understand it anew. They of course also need to absolutely understand technology and all of its potential for good or bad. They must develop a moral framework to lead us all in a world that will need to survive the fruits of our past successes and innovations.

Understanding the Challenge

We have an obvious option – do nothing remarkably different and hope that organic change will naturally lead us into workable global governance. Is this likely? I don't think so. Our 'rise and fall' history evidences that we don't change when necessary; rather we seek change when it's too late. Telling people that 'something is neigh' gets little attention until that something has actually eventuated and by then, it's usually too late to do anything about it.

Organic, adaptive change, managed by linear thinking minds will not be able to deal with the exponential change that technologies are manifesting. Globally we have no concept of how we might govern into the future. This is no startling revelation given that we are hardly governing well today. People still parroting labels such as democracy, socialism or monarchy as models of governance are just not appreciating that these are not actually workable, have led to the current dystopic world and are no answer to the issues we will face individually and collectively as our human paradigm evolves.

It is useful to consider how the largest nation on earth is governing itself and how it conducts its foreign affairs. We need to acknowledge that the larger the population, the larger the diversity of purpose, meaning and telos among its people. China as our most populous nation is worthy of detailed study. Notwithstanding the hypocritical commentary emanating from the west about human rights violations in China, their people exhibit a pragmatic attitude and a wise acknowledgement that personal sacrifices for the common good are a necessary consequence of a large population.

The Chinese are not stupid, nor are they unaware of personal restrictions and imposed compliances; they understand well enough and face reality with the attitude that to me reflects a 'we don't have this or that and we can't do this but that's OK'. They have accepted the wisdom that seems unattainable to our modern western minds. The minds that shout about the freedom of expression, that maintain their right to insult, mock and degrade what is different and what they do not accept or understand. [*These same voices condemn restrictive regimes whilst imposing crippling sanctions and exporting conflict around the world. They are the spoiled, self-arrogated voices of an ignorance born out of an unjustified sense of their way is right; just as their God is the one and only God.*]

It is these people, the confident and self-righteous, the beneficiaries of centuries of exploitation and the heirs to fortunes not earned that will be most shocked by the changes that are coming. They and their once considered great and dominant powers will be exposed and shamed (its already happening).

The near future, dominated by the information age, will change our geopolitical landscape and have significant economic impacts. We will see the transition of institutions born in the industrial age as they adapt to a world previously ordered by advantage trade and the politics of 'military might'. How well this will enable any equal distribution of the benefits of promising abundance technologies and how the globe might achieve any form of reassuring security, remains to be seen.

Looking at how our most powerful nations are governed today is revealing. The happiest people, those that are most satisfied with their leadership – well they're Russian. This exemplifies political leadership. Russia has a very popular leader in President Vladimir Putin. Oh the West doesn't like him much; he's refusing to play their transparent games, consistently points out their hypocrisy and illegal conduct and he has a strategy and a plan for his people. Like it or not, Putin is an outstanding leader and brilliant strategist. As, I said, his example highlights the fact that excellent leadership can unite people, people that will accept hardship for the good of the country, people that can say no to western trinkets and false ideologies, people prepared to do without and make the best out of what they do have. This is only possible because they are inspired, motivated and believe in their leader and therefore themselves.

It might surprise a lot of people that many 'ordinary' Russians are actually wishing to return to communism! Have they already forgotten how bad it was? Have they forgotten free medical care, guaranteed housing and employment and no homelessness? Have they not been swayed by the wonderful inequality, mass unemployment, domestic violence, constant struggles and shiny trinkets that the West promotes? Have they not been convinced that capitalism that is enriching ever fewer elites is good for ordinary people?

Russia is currently benefiting from effective and exemplary leadership. No doubt. This does however also point to a dependency on outstanding individuals and carries the risk of maintaining the quality of leadership. This dependency on individuals requires careful succession planning and a flexible constitution that allows outstanding leaders not to be limited in tenure.

Just as Russia and specifically President Putin and his team, exemplifies excellence in leadership, China exemplifies intelligent and wise governance. Not concerned with labels, China is freeing society at a manageable pace whilst retaining the necessary and yes some might say restrictive framework that enables it to control the largest population on earth. The Chinese government continues to demonstrate deliberate and wise strategic thinking and is able to befriend all, to assist without demand and to pragmatically deal with its own challenges.

I believe that China is showing the way and progressive nations would do well to copy its meritocratic representational governance structure. We need to acknowledge the deliberate training and experience that their government structures facilitate. Their senior politicians typically have over three decades of governance experience. Even mid-level Chinese leaders have governed provinces and managed budgets that are greater than those of entire nations. It must be insulting to their leadership when politicians

HANS J. ORNIG

representing nations of millions or tens of millions offer advice or criticism. In China, these equate to cities or small provincial administrations. We really ought to look to ourselves and learn to keep our unwise opinions in check.

We might all gain from deeply studying and understanding Russian leadership and Chinese wisdom. I appreciate that many will no share my sentiments on these whispered 'enemies' of the West. Nevertheless, I've at least studied leadership and I've experienced a variety of cultures and political systems. I invite you to look again, to look with a beginners mind and to suspend inherited or un-examined beliefs. Look critically at our global situation and consider our future.

I've been privileged to speak with great thinkers and I've had the luxury of contemplation. I'm concerned with our collective future. I see a world that is trivialising itself. I see too many people worrying about the minutia of life and being totally absorbed in irrelevance. It is these people to whom the future will be the greatest shock. When the new era finally dawns on these people, and that is the vast majority of us, they will be lost and looking to leadership. Where will they find it?

Summing Up

Lots of unknowns and they will not resolve organically. Solutions to near future challenges will emerge from various locations and be implemented where political or commercial incentives exist. They will materialise, either in a managed fashion or as is more likely, an ad hoc manner. There are some certainties for the near future – a significant number of people will benefit from the new age technologies; they will take advantage of the very powerful transformational technologies on offer. Many will adapt, will create new organisations, will lead the way and ride the tsunami of change into a spectacular future; the majority, if not led and governed well, will be crushed in its wake.

At every level, exceptional leaders will be required, a new and enhanced type of leader, one competent in current best practice (these are already rare today) and possessing the qualities and skills needed to effectively define and usher in a future. These influencers, leaders, capable of transitioning to a technology evolved mind set are the exponential leaders of the future.

Exponential technologies are propelling us into an uncertain future; a future that will force us to confront anew what it means to be human and the 'meaning of life'. Today, that future is approaching without a plan, no control and no brakes. We are in default leadership mode.

Presumably you thought this book to be about leadership. You might have come to the conclusion that it is actually about behavioural science, technology and obviously, as the title implies - future leadership. It is actually about anthropology; it's about humanity, our collective and individual lives and our ever more joint, interdependent experiences.

Oxford's inaugural professor of anthropology, Sir Edward Burnett Taylor [59], considered the founder of modern anthropology, understood this very well. Writing in the post industrial revolution late nineteenth century, he appreciated the significance, the changes in human behaviour and the impacts on culture that technology drives. His concluding observations in his definitive work 'Primitive Culture: Researches into the development of Mythology, Philosophy, Religion, Language, Art, and Custom' are astoundingly appropriate today and somewhat ominous. I've quoted excerpts from his final paragraph at the beginning of many of the previous chapters but here is the paragraph in full:

> *"It is our happiness to live in one of those eventful periods of intellectual and moral history, when the oft-closed gates of discovery and reform stand open at their widest.* How long these good days may last, we cannot tell. It may be that the increasing power and range of the scientific method, with its stringency of argument and constant check of fact, may start the world on a more steady and continuous course of progress than it has moved on heretofore. But if history is to repeat itself according to precedent, we must look forward to stiffer duller ages of

*traditionalists and commentators, **when the great thinkers
of our time will be appealed to as authorities by men
who slavishly accept their tenets, yet cannot or dare
not follow their methods through better evidence to
higher ends.** In either case, **it is for those among us whose
minds are set on the advancement of civilization, to
make the most of present opportunities,** that even when
in future years progress is arrested, it may be arrested at the
higher level. To the promoters of what is sound and reformers
of what is faulty in modern culture, ethnography has double
help to give. **To impress men's minds with a doctrine
of development, will lead them in all honour to their
ancestors to continue the progressive work of past ages,
to continue it the more vigorously because light has
increased in the world,** and where barbaric hordes groped
blindly, cultured men can often move onward with clear
view. **It is a harsher, and at times even painful, office
of ethnography to expose the remains of crude old
culture which have passed into harmful superstition,
and to mark these out for destruction. Yet this work,
if less genial, is not less urgently needful for the good
of mankind. Thus, active at once in aiding progress
and in removing hindrance, the science of culture is
essentially a reformers science.** (Edward Burnett Taylor)"*

Just like Taylor, we also live at a time of great impact to humanity. We may
not agree that it is 'our happiness' to experience this but perhaps it ought
to be. Just like then, today, great thinkers and leaders, individuals like
Stephen Hawking, Xi Jinping, Ray Kurzweil, Peter Diamandis and John
Demartini are being 'appealed to' to lead thoughts'. Modern industrialists
and entrepreneurs like Robin Li, Elon Musk, Jack Ma and Bill Gates
are today advancing our civilisation and making the most of present
opportunities.

It is institutions like Singularity University that are impressing minds with
the 'doctrine of development'. That is what 'abundance' is all about. The

'light' that has 'increased in the world' is technology. Humanity is indeed building on and continuing the 'progressive work of past ages'.

Taylor lived in the post industrial revolution age. In fact, the industrial age had been going for about a hundred years when he penned his thoughtful conclusion. That greatly transformative period was punctuated by two world wars. Industrialisation created technological advantage and hastened scientific discovery and innovation. It also led to conflict; not the cause, but the ability to execute it globally.

Presently, we are in the early days of the information revolution. We have much to anticipate. Exponential technology is a tsunami about to transform how humanity lives and experiences life. It may be apparent to many that the transformations have begun. There is a changing world order. Inequality, displaced people, suffering caused by disease, famine, warfare, terrorism and the lack of effective global moral and ethical governance is inviting, no, demanding, change.

The last three sentences in Taylor's conclusion are thought provoking. When he writes "...to expose the remains of crude old culture which have passed into harmful superstition..." he writes about reformation (the science of culture is essentially a reformer's science). His phrase 'to mark these out for destruction' (the crude old culture) describes the geopolitical reality about to impact the world order.

What about You?

More and more frequently I read and hear about businesses acknowledging the exponential growth in technology and its wide implications. Rarely however do I hear any such admission from western governments. They still postulate about energy targets, superannuation policy and healthcare as if they are on a different planet; one not burdened with the need to recognise technology implications or the opportunity to control, legislate and steer developments for the common good. At least business leaders are openly admitting that they need to transform or be transformed. What is clear is that the CEOs, the managers, the business leaders that don't get this, will simply not be around much longer.

Understanding the disruption of technological innovation, the many 'Kodak moments' to come, the interconnected global market and the economic challenges ahead, it is not surprising that industry is anticipating and to varying extents, preparing to meet that future. But - what about the individual? What about you?

We are rapidly heading into a dystopian future. Let me spell that out. Whilst we admirably aim to bring about a utopian abundance we are almost certainly in for a very tumultuous time of transition. The world will, for the vast majority, be dystopian – simply 'not a good place'. This is not a pessimistic or even clever prediction. The chapter on global context spells out some of our entrenched ills. Add to that the introduction of revolutionary technology into a world that can demonstrably not manage itself well now and you get the picture.

What can be done about that? We needn't be concerned with business. Whilst businesses will fail in unprecedented numbers, new, exponential organisations and institutions will emerge and flourish. Governments, those still in denial, will self-destruct or be removed. Technology is advancing and governments that have no answers will disappear. Contemplating a future where, based on today's technology, nearly fifty percent of all jobs can be done by robots, machines or simply won't exist, employment will be a key factor. There is the obvious strategy of paying everyone a living salary but when will this start? There are people losing jobs today that are highly unlikely to ever work again. There are young people, even graduates with meaningful qualifications that will never gain paid employment. When will they be paid an income and how will they survive the waiting?

You, I, we all have a choice. Do we embrace the future or hide from it. Ignoring it won't make it go away. The evolutionary divergence has commenced. There are sections of our society that will be left behind. The elderly that today can't set-up their TV or even operate a remote control unit; the people that aren't using mobile phones or the internet and those that just don't want to or can't learn new skills will be the first consigned to the redundancy scrap heap. Then there are those that can't afford to get the latest and greatest applications. Those that aren't 'online' to receive their bills and pay accounts will slowly simply not be catered for. Employment, access to goods and services and basic social interactions will become socially divisive mechanisms.

Fundamentally there is no real choice. If you want to be part of the future in any fulfilling way you need to get on board. Failing to do so, you will still experience change and upheaval; you just won't benefit from it. If you haven't done so yet please reflect on your expectations and your unique telos.

Do you get it? The first and most crucial step in making the most of the future is of course knowledge. The second step – well that's when we decide what to do with it. For now, we need to have the conversation and decide who's to be at the deliberation table.

EPILOGUE

The View from Space

[Please recall the prologue 'The View from Space']

What do you detect?

You are overcome with a deep sense of loss. Your memory is of a blue and green planet; a planet joyfully glistening in its reflective oceans; a unique planet where the universe seems to have conspired to create an idealistic environment for a multitude of life.

As you gaze down at the hot, lifeless and barren planet you hear the echoes of words from humans long gone. "I have a dream" [Martin Luther King, 1963] and "ask not what your country can do for you; ask what you can do for your country" [John F. Kennedy's inaugural address] and "be the change you want to be in the World" [Mahatma Gandhi].

You recall the teachings of Buddha: "All that we are is the result of what we have thought. The mind is everything. What we think we become".

You sadly wonder 'what indeed were they thinking'? What had become of their visionary leaders, hopes and dreams? How did it all go so wrong?

You resolve to continue your exploration of the universe. Still looking for a witness to your own self, you turn away from the place of your creation.

NOTES

1. Rob Nail, the CEO and one of the co-founders of Singularity University.

2. Singularity University (SU) was founded in 2008 by Ray Kurzweil and Peter Diamandis is a California and is situated at NASA's Research Park in the heart of the Silicon Valley, California. It is a benefit corporation with the mission to educate, inspire and empower leaders to apply exponential technologies to address humanity's grand challenges. See also: singularityu.org, twitter.com/singularity and facebook.com/singularity.

3. Salim Ismail, founding Executive Director and Global Ambassador of SU, co-authored (with Yuri Van Geest and Michael Malone) the book 'Exponential Organisations – why new organisations are ten times better, faster, and cheaper than yours (and what to do about it)'; published by Diversion Books, 2014.

4. Skunk works; the term originated from the comic strip Li'l Abner ('Skonk Works' was the name of the moonshine factory) but has since been changed to satisfy copyright. Large corporations, businesses, and innovators are creating task based teams to work outside the constraints of their organisations in order to achieve significant results.

5. Homebrew Computer Club; formed in Silicon Valley, California in 1975 as a gathering of technically minded hobbyists (early nerds?) many of whom went on to play major roles in the IT sector.

6. Dr. Peter H. Diamandis is an international pioneer in the fields of innovation, incentive competitions and the commercial development of space. He is Chairman and CEO of the X PRIZE Foundation, co-founder and Vice-Chairman of Human Longevity Inc. (HLI), co-founder and Executive Chairman of Singularity University,

co-founder/co-Chairman of Planetary Resources Music synthesizer, co-founder of Space Adventures and Zero-Gravity Corporation. He co-authored (with Steven Kotler) two New York Times Bestselling Books: 'Abundance – the future is better than you think' and 'Bold – how to go big, create wealth and impact the world'.

7. Dr Clarence Tan - Adjunct Professor at Bond and Griffith University, former Singularity University Asia Pacific Ambassador, founder of Wireless Applications/Bond Wireless; frequent speaker on exponential technologies and promoter of Singularity University in the Australasian region

8. John C Maxwell. John is one of the world's greatest teachers, mentors and leaders. He has written about 80 books (he's not even sure sometimes!) on leadership and associated topics, was named the Top Leadership and Management Expert in the world by Inc Magazine (May, 2014), ahead of Seth Godin and Jack Welch and has been the consistent number one Leadership Guru on the Global Gurus research organisation site for six years now. [Disclosure – I'm a proud member of the international John Maxwell team (JMT)].

9. WHO, Global Health Observatory data; average life expectancy at birth of the global population in 2013 (see: http://www.who.int/gho/mortality_burden_disease/life_tables/situation_trends/en/

10. James C. Riley (2005) – Estimates of Regional and Global Life Expectancy, 1800–2001; Population and Development Review; Volume 31, Issue 3, pages 537–543, September 2005.)

11. Dr Alan Watkins; TEDxPortmouth, March 2012 'Being Brilliant Every Single Day'

12. David Kriesel, 2007, A Brief Introduction to Neural Networks, available at http://www.dkriesel.com

13. FLOPS (Floating Point Operations per Second) are a standard computer performance measure:

kiloFLOPS or KFLOPS equals 10^3 FLOPS (thousand)
megaFLOPS or MFLOPS equals 10^6 FLOPS (million)
gigaFLOPS or GFLOPS equals 10^9 FLOPS (billion)
teraFLOPS or TFLOPS equals 10^{12} FLOPS (trillion)
petaFLOPS or PFLOPS equals 10^{15} FLOPS (quadrillion)

exaFLOPS or EFLOPS equals 10^{18} FLOPS (quintillion)
zettaFLOPS or ZFLOPS equals 10^{21} FLOPS (sextillion)
yottaFLOPS or YFLOPS equals 10^{24} FLOPS (septillion)

14. Chris F. Westbury; Cognitive Neuropsychologist at University Of Alberta, USA

15. Ray Kurzweil is one of the world's leading thinkers and achievers. He co-founded Singularity University and is currently the Director of Engineering at Google. His bio is at http://singularityu.org/bio/ray-kurzweil/.

16. Telos; Wikipedia definition: (from the Greek τέλος for 'end', 'purpose', or 'goal'); a telos is an end or purpose, in a fairly constrained sense used by philosophers such as Aristotle.

17. Dr John Demartini is the author of 'The Values Factor', founder of the Demartini Institute; leading authority on personal development and human behaviour.

18. Phrase used by Yuval Noah Harari on TED Global 7/2015

19. Current World Population, refer http://www.worldometers.info/world-population/

20. Socrates (in Plato's 'Phaedrus')

21. Based on a Statista survey – issues most concerning Americans, [https://www.statista.com/chart/4535/americans-say-terrorism-is-the-top-threat-to-their-nation/]

22. Bill Gates presentation at TED2015 titled: 'The next outbreak? We're not ready'.

23. Marc Goodman; Author of 'Future Crimes – everything is connected, everyone is vulnerable and what we can do about it'.

24. US Defence Secretary Donald Rumsfeld; The MSNBC Daily article "Building momentum for regime change: Rumsfeld's "secret memos" (24) (02/16/13 09:27 AM—UPDATED 02/27/14 05:05)

25. Chinese foreign ministry spokesman, Lu Kang quoted in the article "China urges efforts to look for solutions, not trouble on AIIB" as published by Xinhua, 18 June 2015.

26. Attorney Ellen Brown (Global Research, 22 June 2015)

27. Article in British Guardian Newspaper, May 2015

28. Foreign Affairs, September/October 2014

29. See www.shadowststs.com

30. Refer www.commonwealthfund

31. See www.census.gov

32. UNICEF and WHO report 'Progress on Sanitation and Drinking Water: 2015 Update and MDG Assessment'

33. Global Trends Report, Office of the UN High Commissioner for Refugees, 2015

34. Reuters report dated 19 Jan 2015

35. Diamond, L.; lecture at Hilla University for Humanistic Studies January 21, 2004: 'What is Democracy?'

36. Ray Kurzweil; published biography at http://singularityu.org/bio/ray-kurzweil/

37. Deloitte's sixth Technology Trends Report (Tech Trends, 2015: The fusion of business and IT)

38. Sean A. Hays' 2011 post at www.idgconnect.com

39. Ray Kurzweil's 'Law of Accelerating Returns'; March 7, 2001

40. Alan Murray; 15 July 2015, Fortune CEO Daily, fortunedaily@newsletters.fortune.com [http://fortune.com/2015/07/15/ceo-daily-wednesday-july-15/]

41. Sensors listed at https://en.wikipedia.org/wiki/List_of_sensors

42. Swiss referendum on an unconditional basic income; reported by Deutsche Welle, 1 Feb 2016

43. Categorisation of robots based on the International Federation of Robotics; http://www.ifr.org/

44. Aaron Saenzon; Singularity Hub article 'We Live in a Jungle of Artificial Intelligence that will Spawn Sentience', August 2010

45. Cognitive Assistant that Learns and Organizes explained at https://en.wikipedia.org/wiki/CALO

46. Computer writing - http://www.theguardian.com/technology/2015/jun/28/computer-writing-journalism-artificial-intelligence

47. Chinese stem cell research (iPS) at http://english.cas.cn/head/201505

48. Andrew Ng; quoted in Bloomberg report October, 2014

49. Robin Li Yanhong, Baidu's founder and CEO; quoted in the article 'Baidu chief seeks PLA backing for AI project', South China Morning Post, 3 Mar 2015.

50. China's Military Strategy, released in Beijing, May 2015 by the State Council Information Office of the People's Republic of China (China Central Television News - http://english.cntv.cn/2015/05/26)

51. Huawei; www1.huawei.com

52. Europe's Human Brain Project; https://www.humanbrainproject.eu/

53. Japanese Research Organisations Contribute to Human Brain Project, Okinawa Institute of Science and Technology Graduate University News, 29 Jan 2013

54. Singapore's BCI projects detailed at http://nsp.i2r.a-star.edu.sg/projects

55. Iichi Lee, President of Korea's Institute of Brain Science www.brainkibs.org

56. MIT Review 23 Feb 16

57. Gale Smith; gsmith@houstonmethodist.org

58. Abstract from: http://www.nature.com/nbt/journal/vaop/ncurrent/full/nbt.3506.html

59. Sir Edward Burnett Taylor; Oxford's inaugural professor of anthropology, (1832 – 1917) definitive work 'Primitive Culture: Researches into the development of Mythology, Philosophy, Religion, Language, Art, and Custom' (first published in 1871);

60. Citi GPS: Global Perspectives & Solutions, January 2016

61. 'Bold – how to go big, create wealth and impact the world'; Peter Diamandis and Steven Kotler.

62. Bronnie Ware, author of the book 'The top five regrets of the dying'.

63. J. Mariah Brown; article Servant Leadership: Leading the Way Through Servitude

64. Klout Score; www.klout.com

65. Pareto principle'; explained at: http://betterexplained.com/articles/understanding-the-pareto-principle-the-8020-rule/.

Printed in the United States
By Bookmasters